roadside geology

of northern california

David D. Alt
Donald W. Hyndman

MOUNTAIN PRESS PUBLISHING CO.
Missoula, Montana

16th Printing, March 1994

Library of Congress Catalog Card Number: 74-81834

ISBN 0-87842-055-X

MOUNTAIN PRESS PUBLISHING COMPANY
P.O. Box 2399
Missoula, MT 59806
(406) 728-1900

preface

All geologists are plagued by friends staggering under the weight of rocks they want identified. Knowing people who find funny rocks while out picnicking seems to be one of the natural hazards of our profession. We wrote this book for those friends who want to learn a bit about the geologic foundations of their surroundings and did our best to be informative without becoming indigestible. Limiting ourselves to aspects of northern California geology that we thought most people would find interesting, we avoided the more rarefied topics that only geologists enjoy. We did our best to avoid crossing the delicate line that separates simplification from oversimplification.

All geologists begin to identify a strange rock by asking where it came from. Each region has its own geologic style that permits some rocks to form and prohibits many others. We try to point out the possibilities and simplicity within each region to give a general idea of what to expect. And we also try to convey through words and pictures an impression of the usual appearance of the common rocks within each region. Hopefully you will be able to tell for yourself what kind of rocks you have and how it got to be what and where it is.

Northern California contains about as many geologic complications as nature could fit into such a small area. Until recently it would have been hopeless even to attempt such a brief discussion of the region but geology has changed in recent years and now we begin to see the forest where before we only knew the trees. We did our best to sketch the broad outlines of the overall story as nearly as they are known, and then fit a few of the more interesting features into their proper place in the picture. Recent progress in geologic thought makes this possible by greatly simplifying many things that had seemed terribly complicated.

We have enjoyed working on this book, reading about the geology of northern California and visiting the rocks. We owe a debt of gratitude to nearly every geologist who has written about northern California and regret that we must thank them all collectively because they are too numerous for more personal acknowledgment.

David D. Alt and Donald W. Hyndman
Department of Geology
University of Montana
Missoula, Montana

contents

index to highways

geologic

SYMBOLS		NAME
Lake sediments Glacial debris		*Pleistocene*
	BEGAN ABOUT 3 MILLION YEARS AGO	
Valley-fill sediments	TERTIARY	*Pliocene*
	ABOUT 11 MILLION YEARS AGO	
Flood basalt flows Fragmental volcanic rocks	TERTIARY	*Miocene*
	ABOUT 25 MILLION YEARS AGO	
	TERTIARY	*Oligocene* *Eocene*
	BEGAN ABOUT 60 MILLION YEARS AGO	
Great Valley Sequence Franciscan	MESOZOIC ERA	*Cretaceous:* late middle early
Granite intrusions	MESOZOIC ERA	*Jurassic*
Metamorphosed sedimentary rocks, slate, schist Metamorphosed volcanic rocks	MESOZOIC ERA	*Triassic*
	BEGAN ABOUT 225 MILLION YEARS AGO	
Metamorphosed sedimentary rocks, slate, schist	PALEOZOIC ERA	
	BEGAN ABOUT 600 MILLION YEARS AGO	
Fault: arrows indicate direction of movement	PRECAMBRIAN ERA	

time scale

IMPORTANT EVENTS IN NORTHERN CALIFORNIA

Ice ages, big lakes in Modoc Plateau, Cascade volcanoes

Sierra Nevada rises, Sutter Buttes erupts
Valley-fill sediments deposited

Eruption of flood basalts in Modoc Plateau
San Andreas fault begins moving

End of stuffing an ocean floor under edge of continent
Streams of Sierras deposit gold-bearing gravels

Continued stuffing of Franciscan sediments against edge of continent to form Coast Range
Separation of Klamaths and Sierras into separate ranges
Formation of Great Valley; begin filling with sediments
Melting and intrusion of huge granite batholiths in Klamaths and Sierras, formation of metamorphic rocks

Begin stuffing ocean floor under edge of continent to form rocks of Klamaths and Sierras

---

Deposition of old sediments of continental shelf in area of Klamaths and Sierras

---

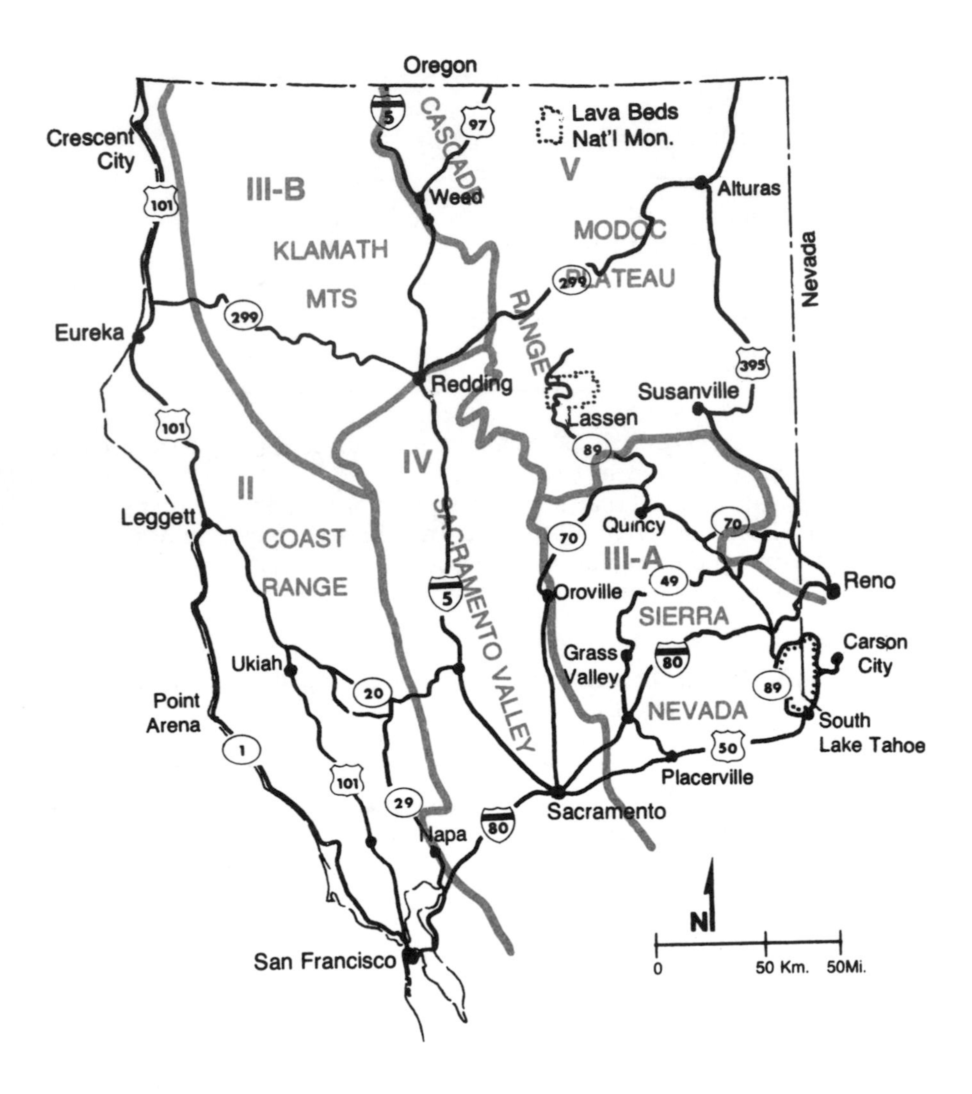

Oregon
Lava Beds
Nat'l Mon.
Crescent City
III-B
CASCADE
V
Weed
Alturas
KLAMATH
MTS
MODOC
PLATEAU
Nevada
RANGE
Eureka
Redding
Susanville
Lassen
IV
II
Leggett
COAST
RANGE
SACRAMENTO VALLEY
Quincy
III-A
Reno
Oroville
SIERRA
Grass Valley
Carson City
Ukiah
NEVADA
Point Arena
South Lake Tahoe
Placerville
Sacramento
Napa
San Francisco
N
0
50 Km.
50Mi.

I

the great collision

— a moving sea floor

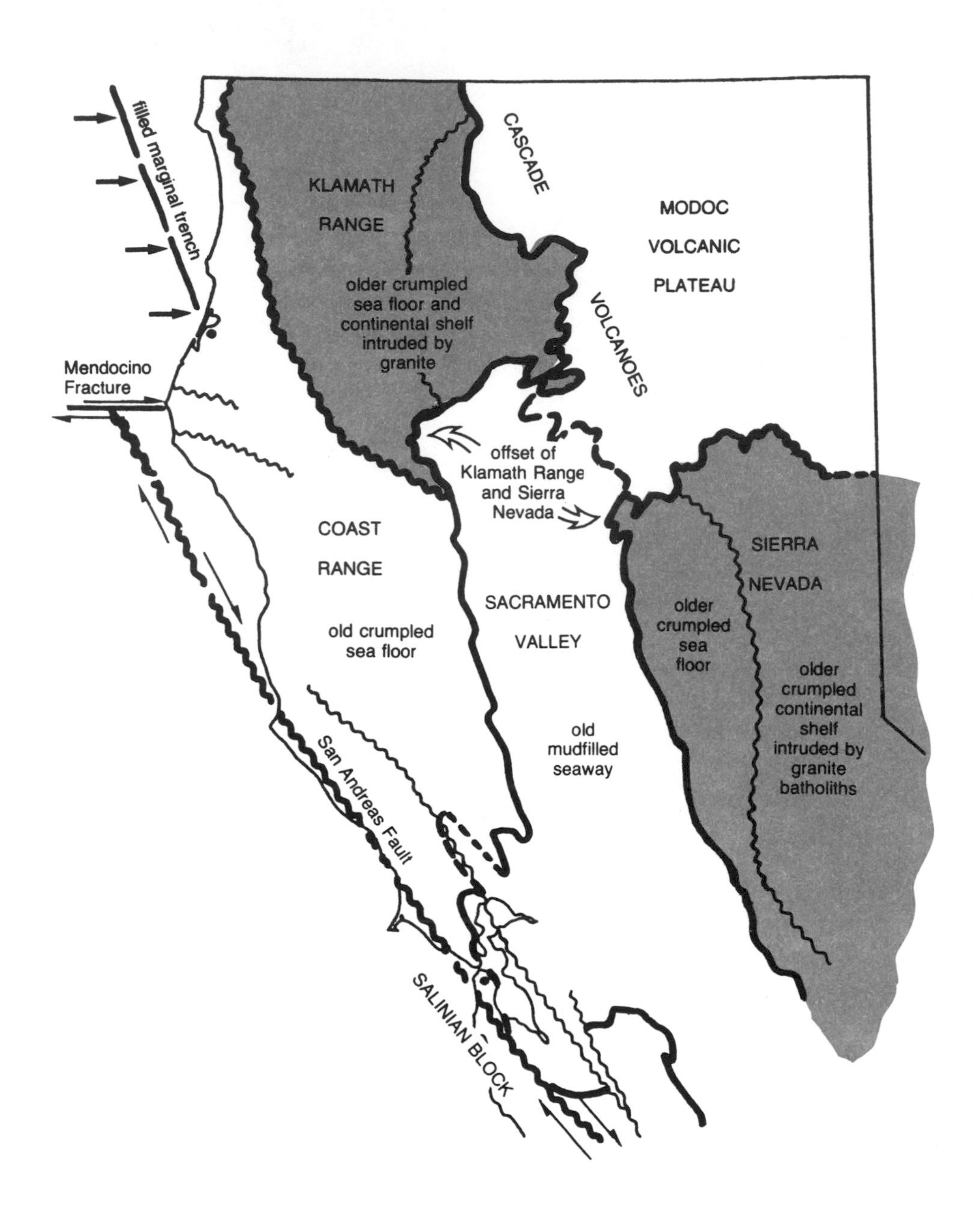
filled marginal trench
KLAMATH
RANGE
older crumpled
sea floor and
continental shelf
intruded by
granite
CASCADE
VOLCANOES
MODOC
VOLCANIC
PLATEAU
Mendocino
Fracture
offset of
Klamath Range
and Sierra
Nevada
COAST
RANGE
old crumpled
sea floor
SACRAMENTO
VALLEY
old
mudfilled
seaway
SIERRA
NEVADA
older
crumpled
sea
floor
older
crumpled
continental
shelf
intruded by
granite
batholiths
San Andreas Fault
SALINIAN BLOCK

the great collision

California is the product of a prolonged head-on collision between the leading western edge of North America and the floor of the Pacific Ocean. It consists of rocks from the deep ocean bottom and mud scraped off them as the continent overrode the ocean basin. The story of rocks from deep in the earth, the muddy sediment that covered them, the crumpling collision that destroyed them both, and the birth of new rocks that resulted, is the story of California's rocks. The details are complicated but the broad picture is not. It is a picture that anyone can see in the rocks of the state.

Until the 1960's geology was a landbound science practiced by people who whacked at outcrops with hammers and thought of the oceans as big wet places where no rocks were visible. They missed some of the best rocks. Modern oceanographic geology, practiced with instruments carried on ships, has shown that the ocean floors are not the featureless plains everyone had long imagined. Instead, they are rugged and actively moving like giant conveyor belts, at a rate of about two inches a year. It is the sea floors, not the continents, that most directly express the internal processes that shape the earth's surface.

Continents are really rafts of light rocks, granite, for the most part, floating embedded in the heavy black rocks of the earth's mantle — the rocks that make the bedrock sea floor. Continents float on the surface of the earth as though they were mats of foam drifting in the eddies of a stream or scraps of milk skin floating on the bubbling surface of a pan of hot cocoa. The submerged topography of the sea floor, the outer surface of the earth's mantle, reflects the restless movements of the earth's interior that sweep the continents along.

If the water could somehow be drained from an ocean, we would see revealed a flat basin with a broad ridge winding down its middle and lengthy deep trenches along the edges near the continents. Following the crest of the mid-ocean ridge there would be a series of deep cracks. Bare, black bedrock would be exposed near the middle of the ocean on the gentle slopes of the mid-ocean ridge; closer to shore the sea floor would be blanketed with thick deposits of muddy sediment shed from the continent.

Common black volcanic basalt is almost the only hard rock exposed on the sea floor. This seemed terribly monotonous until rock ages determined by analysis of radioactive minerals showed that the basalt near the mid-ocean ridge is brand new and that near the edges of the oceans may approach 200 million years old. Since sea floor is always youngest at the mid-ocean ridge and becomes older near the edges, we must conclude that new sea floor is forming at the mid-ocean ridge and moving away from it towards the edges of the oceans.

If new sea floor is forming, and if the earth is to remain the same size, then old sea floor must be disappearing. The deep trenches around the margins of the ocean basins are the places where old sea floor is sinking into the hot depths of the earth's mantle. Evidently the trenches are pulled down by the descending flow of rock just as a dimple forms in the water surface when a drain is opened beneath.

Most geologists believe that there are great currents in the earth's mantle driven by heat released from decay of radioactive elements. The same kind of currents, generally, that move in a kettle of soup heating on the stove or in a cooling cup of coffee. Heat slowly accumulates in the rocks deep within the earth, expanding and partially melting them so they become less dense than the rocks above. Then the hot rocks rise slowly through the earth, breaking the surface at the mid-ocean ridge. There they begin to cool and to spread slowly away in both directions from the crack in the ridge, pushed along by more hot rock rising from beneath. They continue to cool during their long sojourn across the ocean floor and finally become dense enough to sink back into the interior creating an ocean deep. So the entire ocean floor is really a sort of giant conveyor belt rising from its source in the mid-ocean ridge and sinking as a great slab to its destiny in the deep ocean trench. This great planetary conveyor belt seems to move about two inches per year, ponderously slow in human terms but unseemly haste for a geologic process. Movement at that rate can carry the sea floor from its source in mid-ocean to the edge of the widest ocean in less than 200 million years and nowhere is sea floor known to be older than that.

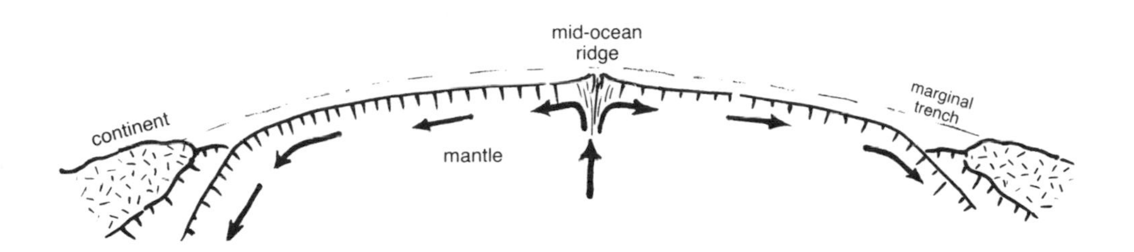

The sea floor does not move in one piece but in a series of segments that behave almost as though they were rigid. These slide past each other along great faults and where they carry the continent along with them they tear it into large slabs. One of these large sea floor faults, called the Mendocino fracture, projects across northern California along the line of the southern boundaries of the Klamath Mountains and Modoc Plateau. The San Andreas fault is another such boundary between slabs of continent riding on two moving segments of the mantle.

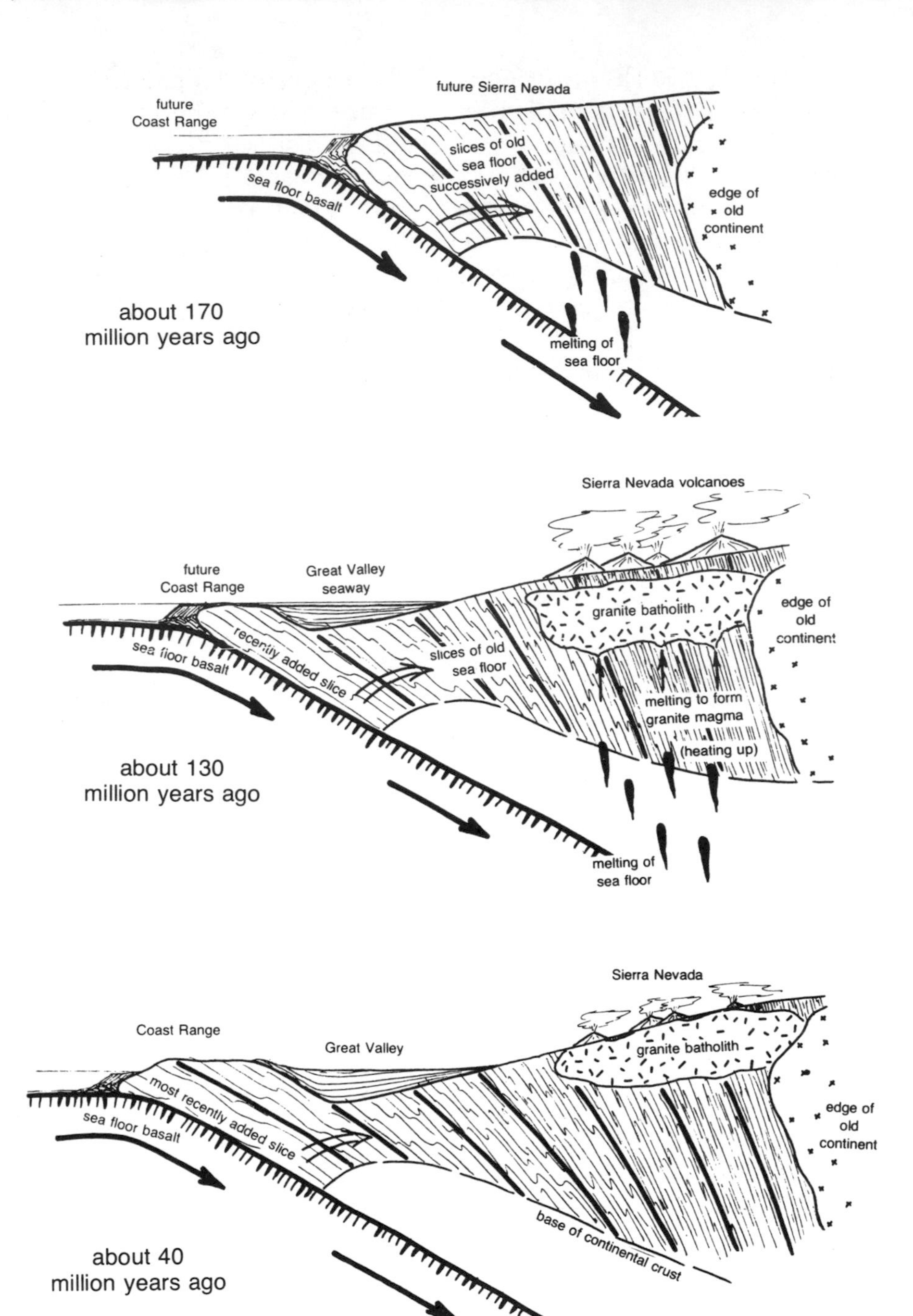

Stages in the growth of California, a recent addition to the North American continent.

So long as the continent rides along on the mantle as though the two were solidly glued together there is relatively little geologic action. The slow processes of erosion gradually reduce the stable land to low plains while rivers dump the eroded soil into the ocean where it accumulates as thick deposits of mud along the edge of the continent. The east coast of North America is that way now and the west coast was for a long time before about 200 million years ago. Then the continent and mantle became uncoupled along the west coast to begin the sequence of geologic events that created California.

Where the continent and mantle are not coupled together, the moving rocks of the sea floor slide into the marginal trench, scraping off their burden of muddy sediments against the edge of the continent. These form coastal mountain ranges. Underneath them the dark rocks of the sea floor ride on beneath the continent where they begin to melt forming magmas that rise through the crust to feed chains of volcanoes parallel to the coast. Thus the sediments dumped on the sea floor are returned to the continents creating new ranges of mountains that will in time be eroded and returned again to the ocean. Were it not for this continuous process of renewal, the processes of erosion would long ago have reduced the surface of the earth to a featureless, watery swamp.

When the moving floor of the Pacific Ocean began to slide beneath the edge of the continent, thick deposits of sediment near shore were crushed together and jammed onto the old continent to make the crumpled metamorphic rocks of the Sierra Nevada and Klamath Mountains, the wreck of what was once a quiet coastal plain. Generous slices of the black sea floor itself were incorporated within them to become the broad belts of dark rocks that lace through both ranges.

As the action continued, the growing welt of new coastal mountains built steadily westward as the moving sea floor stuffed slice after slice of muddy sediment under its seaward edge. And each new slice of sediment was brought in from farther out at sea. So the rocks in California tend to become younger westward and to include sediments deposited farther from shore.

While the younger slices of deep sea sediment were being stuffed into a marginal trench to make the rocks that later became the Coast Range, the coastal mountains already formed were torn into two segments that moved about 60 miles apart to become the present Sierra Nevada and Klamath Mountains. Although no one can be sure, it seems likely that the Klamath Mountain block moved west opening a seaway gap behind it in the northeastern part of California, that part destined to later become the Modoc Plateau.

Also while the younger sediments were accumulating to become the Coast Range, molten magma rose from the descending slab of sea floor into the Sierra Nevada and Klamath Mountain rocks that had already formed. Long chains of volcanoes rose above them and part of the older sedimentary rocks also melted. Some of the new magma erupted at the surface but most of it solidified within the crust to become the enormous masses of granite that we see in those mountains today now that they are deeply exposed by erosion. The tremendous quantity of heat in the crumpled sedimentary rocks cooked them and welded them into solid metamorphic rocks. Because the Coast Range rocks were the last swept together before the action stopped, they were never intruded by large masses of molten magma and never so completely welded by recrystallization.

It seems likely that rocks of the Coast Range were stuffed into a deep marginal trench offshore that was then being pulled down by the slab of sea floor sinking beneath it. After the action had stopped — these slow currents in the mantle shift about and change positions every hundred million or so years — the trench no longer had anything pulling it down and so it floated up. It was then that the crumpled rocks of the Coast Range that had been packed into the deep marginal trench broke the surface as a chain of islands. Their rise separated the future Great Valley from the main body of the Pacific Ocean creating an isolated inland sea that gradually filled with sediment.

It is hard to know what happened in California after the last Franciscan rocks had been stuffed into the marginal trench that was destined to become the Coast Range. It is difficult to believe that all geologic activity ceased for 30 million years because the internal earth movements that created California must have continued. Nevertheless, we have very little record of whatever may have happened during the 30 million years after the Coast Range first appeared; it does seem that this may have been a period of relative geologic peace. Whatever else may have happened, we can be sure that the slow processes of erosion removed the volcanoes and reduced the Sierra Nevada to a province of low, rolling hills barely above sea level. The Great Valley and the Modoc seaway in the northeast both filled completely with sediment eroded from the surrounding hills and became dry land.

Instability resumed about 30 million years ago with uplift and stretching of a large portion of the western part of our continent. Action began in California with eruption of a series of enormous basalt lava flows. Much like those normally associated with the mid-ocean rise, they spread onto the area of the former Modoc seaway. These converted what had been a level plain similar to the present Great Valley into a high volcanic plateau several thousand feet above sea level.

Uplift and stretching of the continental crust broke it into a series of great blocks that moved like sections of a concrete sidewalk set on unstable ground. This involved all of the eastern Sierra Nevada and the Modoc Plateau. Movement of the blocks was accompanied by renewed volcanic activity, eruption of smaller quantities of black basalt and extensive blankets of white volcanic ash. The blocks that moved up are now mountain ranges, the largest of them is the Sierra Nevada, and those that moved down became large basins like the Sierra Valley east of the northern Sierras.

While the continental crust east of the Great Valley has been stretching and moving vertically to make mountains and basins, other crustal movements have been rearranging the geography of the Coast Range. A large piece of continent,

Granites in the Coast Range were originally part of the Sierra Nevada and have been carried northward by movement of the Salinian block.

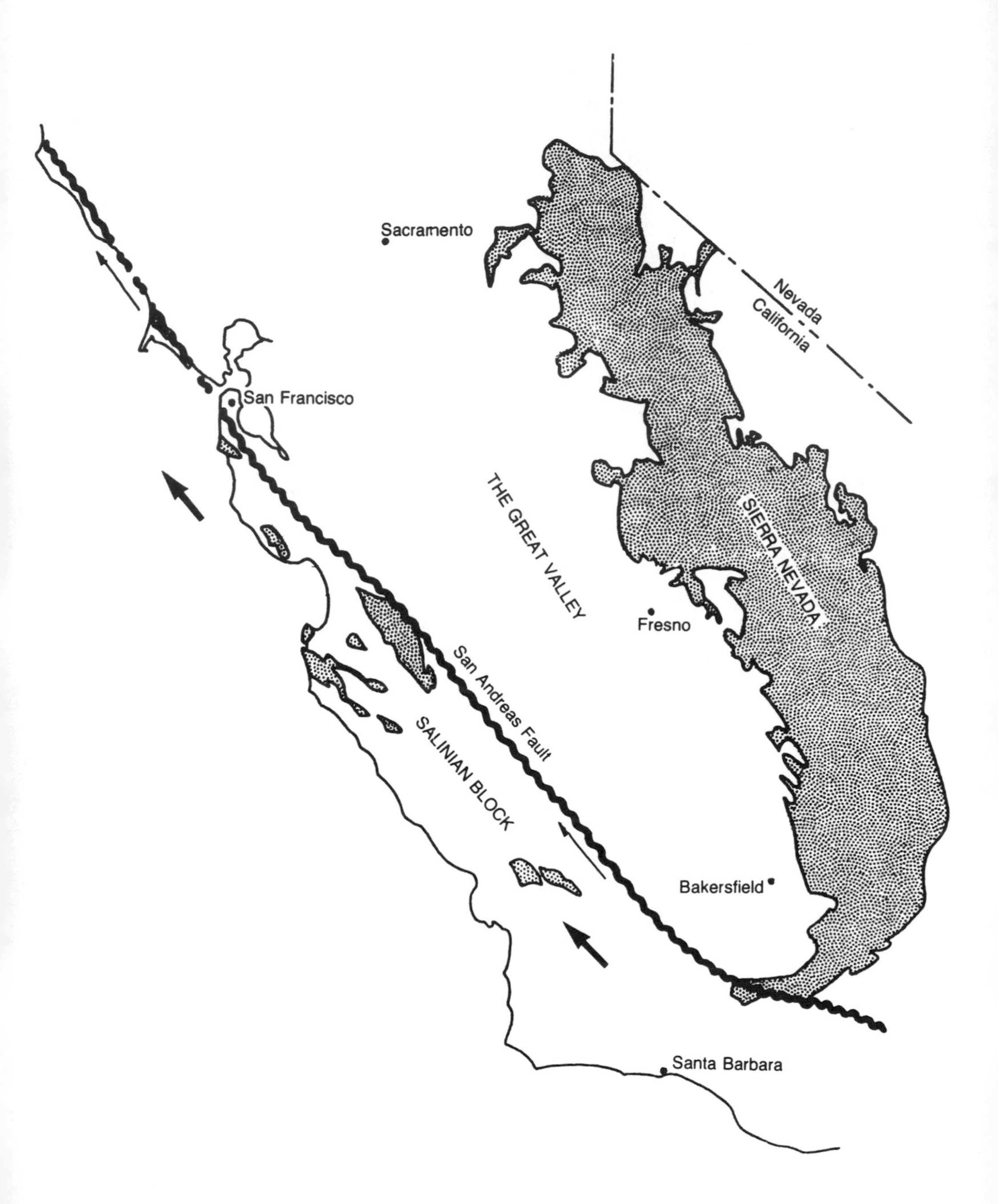

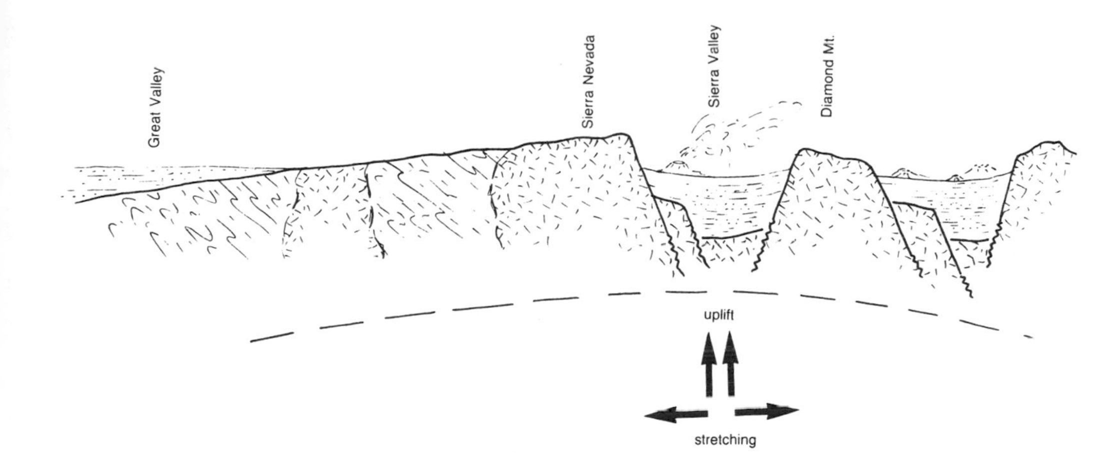

essentially the southern end of the Sierra Nevada, once in southern California, is moving northward west of the San Andreas fault. Evidently it is attached to a northward-moving segment of the earth's mantle.

During the last several million years, very recently as geologic events go, the big Cascade volcanoes have built a curving chain of peaks extending from the broken northern end of the Sierra Nevada to the Canadian border. Evidently another descending current of sea floor off the coast of the Pacific Northwest is now carrying the sea floor beneath the continent where it melts to form magmas. Almost surely, the Cascades are today as the Sierra Nevada was over 100 million years ago. So in California we see all stages of the processes that convert muddy sediments of the sea floor into continental rocks which rise as mountains, erode and return to the sea as sediments, and then are swept back onto the continents again by the constant motion of the sea floor conveyor belt.

II

the coast ranges

— a nightmare of rocks

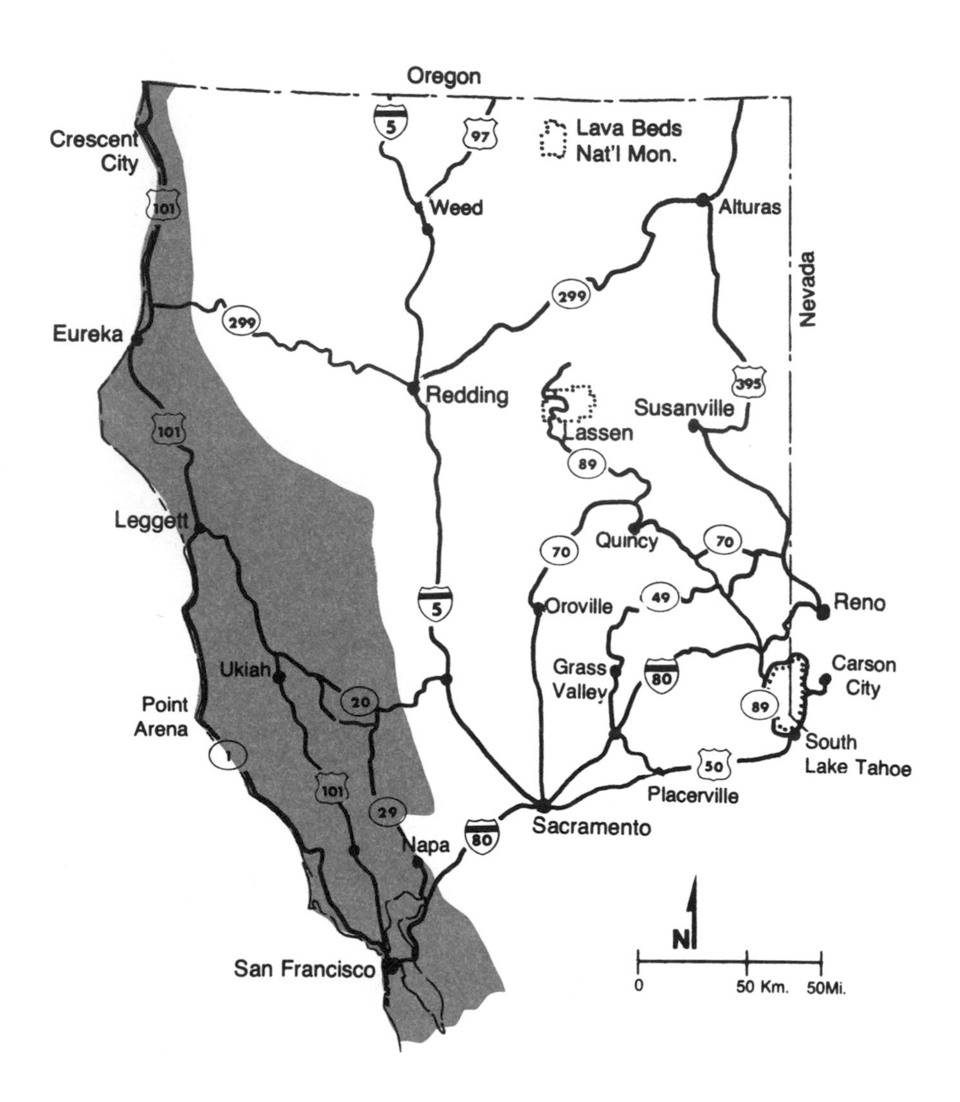
Oregon
Crescent City
Lava Beds Nat'l Mon.
Weed
Alturas
Nevada
Eureka
Redding
Susanville
Lassen
Leggett
Quincy
Reno
Oroville
Grass Valley
Carson City
Ukiah
Point Arena
South Lake Tahoe
Placerville
Sacramento
Napa
San Francisco
N
0
50 Km.
50Mi.
101
5
97
299
395
89
70
49
80
20
1
29
50

— a nightmare of rocks

Generations of geologists have been driven to despair by the scrambled rocks of the Coast Ranges. Rocks that are both complex and nondescript, that clearly defy some of the most basic and dependable principles of geology, that lack vital kinds of evidence and — to add the cruelest insult of all — are very poorly exposed. If all rocks resembled these, the science of geology could never have been developed. Geologists have needed every insight they could gain from study of simpler rocks elsewhere to reach some broad understanding of the chaotic jumble of broken and disordered rocks that are the Coast Ranges.

Despite the chaos, certain broad simplicities have been apparent in Coast Range geology almost since the beginning of serious study. There are four major rock units, always arranged in essentially the same way: The most important of these is the Franciscan sediments, a jumbled mess of muddy sandstones and cherts interlayered with basalt lava flows, the entire assemblage so thoroughly folded and sheared that some large outcrops look as though they have been stirred with a stick. On top of the Franciscan sediments along the east flank of the Coast Ranges is the Great Valley sequence, an assemblage of sedimentary sandstones, cherts, and basalt lava flows, that look about like the Franciscan rocks must have before they were folded. Between the Franciscan sediments and the Great Valley sequence there is a zone a mile or more thick of heavy black igneous rocks and green serpentinites, rocks that seem completely out of place on the continent. The fourth major unit is also sandstone and mudstone, usually full of fossils and clearly much younger than the other rocks of the Coast Range. These rocks do not pose major problems because they are clearly on top of, rather than part of, the Coast Range nightmare.

Franciscan sedimentary rocks and Great Valley Sequence rocks both appear to have begun as sediments on the floor of the ocean, exactly the same kinds of sediments being deposited offshore today. Most of these sediments are muddy sandstones, some so dark they are almost black, interlayered with a few lava flows of black basalt and beds of chert, a rock composed of the quartz skeletons of microscopic animals all welded together byrecrystallization.Some Franciscan cherts are white but most are red, green, or brown. Pebbles eroded from them make gaily colorful gravels in the streams and beaches of the Coast Range. Layered beds of chert like those in the Coast Range are deposited on the deep mid-ocean bottom far offshore beyond the areas where muds and sands from the continent are laid down.

Both the Franciscan sediments and the Great Valley Sequence are many thousands of feet thick and quite monotonous, containing the same kinds of rock throughout. Geologists find it extremely difficult in the Great Valley Sequence, and absolutely impossible in the Franciscan, to recognize individual layers of rock from one outcrop to another. This is frustrating because the usual way of studying sedimentary rocks is to follow individual layers finding out where they go and how they change. There is no better way to trace folds and locate faults, to unravel the rocks and predict locations of such interesting things as mineral deposits.

Geologists usually rely on fossils to place rocks in their correct sequence. Unfortunately, Franciscan and Great Valley sequence rocks contain very few fossils, mostly not very helpful kinds and those in the Franciscan sediments are generally crushed and sheared almost beyond recognition. There is no hope that enough fossils will ever be found to enable geologists to trace the individual rock layers as precisely as they would like. The Coast Range will always protect some of its geologic secrets.

Persistent search over many years did finally turn up enough fossils to solidly prove one amazing fact: The Franciscan sediments and Great Valley Sequence are the same age; both were laid down on the floor of the ocean as deposits of sand and mud between about 150 and 100 million years ago. Experience in other areas had led geologists to expect that the undeformed sediments of the Great Valley Sequence would be much younger than the mangled rocks of the Franciscan — laid down on top of them after the period of deformation had safely ended. How then did the Great Valley Sequence wind up on top of the Franciscan and how did it nearly escape deformation in the process?

Geologists faced with a problem in determining the relationship between two bodies of rock normally begin work by examining the surface of contact between them. This usually reveals whether the rocks are next to each other because they were formed that way originally or because of some later movements in the earth's crust. Even though the Franciscan and Great Valley sequence rocks are always associated with each other, they are not actually in contact anywhere — that broad belt of serpentinite separates them everywhere. So the problem of how the Great Valley Sequence got on top of the Franciscan is compounded by the mystery of how that nasty-looking green serpentinite came between them. The serpentinite is quite a problem itself.

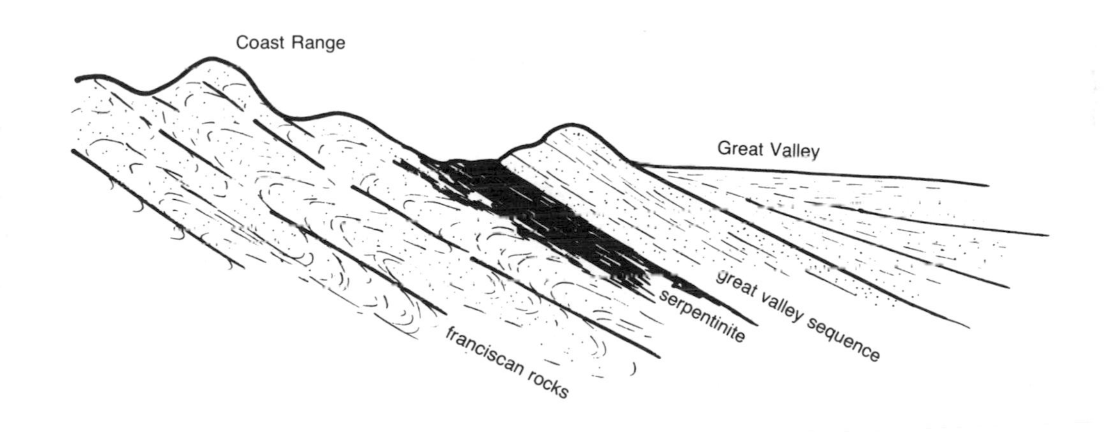

Serpentinite is one of the strangest of rocks. It is always green, occasionally very pale green but most often dark and mottled by patches of black. Invariably it is seamed by a criss-crossing pattern of closely-spaced fractures that have polished surfaces, evidently because they were rubbed smooth by internal friction as the whole body of rock moved. Some serpentinite, especially the lighter-colored varieties, has a soapy feel and is soft enough to whittle with a pocket knife. This is the kind that is often called "soapstone." But the rock is generally so fractured that solid pieces even big enough to make into a small ashtray are hard to find.

How serpentinite forms, where it comes from, and how it manages to get where we find it have been questions for lively debate among generations of geologists. Some of them become quite emotional about the subject. Serpentinite has a chemical composition that strongly suggests an origin somewhere in the earth's mantle beneath the continental crust. Bodies of serpentinite often cut boldly through the enclosing rocks as though they had been forcibly intruded into them as molten magmas. But nothing else about serpentinite suggests that it was ever molten; the arrangement of mineral grains is all wrong for an igneous rock and nearby rocks are never cooked. Because serpentinite has a very high melting point, a molten serpentinite magma, if there were such a thing, would surely cook the surrounding rocks. Many of the minerals in serpentinite contain molecules of water locked in their crystals, something that would be impossible if the rock had formed by cooling of an extremely hot magma.

There are more complications. Coast Range serpentinites often contain angular fragments of a rare, beautiful and perplexing rock called "blueschist." Chunks of it that may be as small as acorns or as large as a county courthouse litter the ground here and there in places where the serpentinites outcrop. These blueschist knockers are easy to recognize because they are heavy, bluish-black rocks liberally flecked with intensely blue crystals.

Minerals in blueschist rocks form under extremely high pressures and rather low temperatures, the kind of pressures that prevail near the base of the continental crust but the kind of temperatures that exist near the surface. There seemed to be no place where these conditions could exist simultaneously. For a long time many geologists felt that these facts were hopelessly contradictory.

Jade is another unusual rock widely distributed in serpentinites. Like the blueschist, it forms under extremely high pressures and chunks sometimes litter the ground in places where serpentinites outcrop. Jade comes in the same colors as serpentinite so it is very difficult to find in solid outcrops; it usually seems to occur in seamy zones where intense shearing has ground the rock fine. Jade is much easier to find in streams or beach gravels because it is hard and tough and survives as rounded pebbles long after the soft serpentinite has been ground away. Any stream draining areas where serpentinites outcrop or any beach near a place where waves erode serpentinite is a likely spot to look for pebbles of jade. Pebbles of green chert eroded from the Franciscan rocks can be very deceiving but jade is noticeably heavier in the hand and much more difficult to break. If a tap with a hammer smashes the pebble, it was chert. Jade is very tough and small pebbles will easily withstand a hammer blow that would drive a light nail.

Mercury mines and serpentinite belts also go naturally together. The ore consists of blood-red cinnabar disseminated through the serpentinite and adjacent rocks, mostly coated on fracture surfaces like paint. Simple roasting liberates the mercury as vapor that can be converted to the liquid metal by cooling it in a condenser. Mercury mining began very early in California because the metal was used for recovering gold from gravels washed in pans or sluices. Some of the methods used then to roast the ore and condense the vapor, which is extremely poisonous, were shockingly careless and must have poisoned the air for miles around. Even though techniques have improved greatly since then, mercury mining and refining still pose serious environmental threats and require very careful operation. The economics of mercury mining seem to be nearly as volatile as the metal so most of the mines have a long record of intermittent operation.

Manganese is a scarce and vital mineral resource that occurs in deep sea sediments both on the ocean bottom offshore and within the Coast Range. Numerous small deposits in the California Coast Range were mined during the two world wars when normal supplies failed and prices were high but none of them are large enough to supply a large mine. Most of the manganese minerals are black and easy to overlook because they look like stains on rock surfaces or fill inconspicuous fractures. Other manganese minerals, kinds rarely seen except in mines, are an unforgettable shade of hot pink.

Geologists finally realized, as they learned something about the oceans, that the Coast Range is made of sediment scraped off the floor of the Pacific Ocean. Franciscan sediments and Great Valley Sequence rocks are indeed the same, just as the fossils and rocks suggest, except that the Franciscan rocks were jammed onto the edge of the continent while the Great Valley Sequence rode undisturbed above them. Slice after slice of muddy sediment was scraped off the sea floor as it disappeared beneath the continent, at a rate of about 2 inches per year. One under the other, the slices were stuffed onto the continental margin to make the swollen welt of Franciscan rocks that is the Coast Range.

Like all genuinely good ideas in science, the notion that Franciscan rocks are simply sea floor sediments scraped off against the continental margin simplified things that had seemed complex. The sediments must have been brutally deformed as they were jammed onto the edge of the continent so it is not surprising that individual layers can hardly ever be followed from one exposure to another. In fact, the present arrangement of the layers must have very little relationship to their original arrangement on the sea floor. Layers of muddy sandstone and chert now exposed next to each other in the same roadcut may well have been deposited dozens of miles and millions of years apart, and the worst suspicions of some of the geologists who have tried to sort out the Franciscan are amply confirmed.

Even the enigmatic green serpentinites with their strange chunks of blueschist make sense. Serpentinite is very similar in composition to the rocks beneath the continents and the sea floors except that the serpentinite appears to have been altered as it absorbed water, presumably from the wet sediments. That broad belt of serpentinite separating Franciscan from Great Valley sequence rocks is nothing more than a generous slice of the bedrock seafloor that exists everywhere beneath the surficial veneer of muddy sediments — in this case beneath the Great Valley Sequence.

Everything we know about serpentinite suggests that it is a very mobile rock easily deformed by squeezing. Slices of it caught in the Franciscan schists as they were crushed against the continent seem to have squirted through the mass like globs of grease caught in a bale of scrap. Stray masses of serpentinite squeezed through the compacting mass of Franciscan sediments penetrating wherever they found a way. Some were swept deep beneath the crust before they escaped to squeeze back towards the surface; these are the masses that now contain chunks of blueschist formed during the brief sojourn at depth while the rock was subjected to very high pressures but did not remain there long enough to get very hot. All the serpentinites display evidence of their mobility in their numerous shinny fracture surfaces polished by internal movement within the mass.

Evidence obtained from study of the few surviving fossils and from analysis of radioactive minerals in the rocks shows that the Franciscan sediments were crammed against the continent between 150 and 100 million years ago. During the same time enormous masses of molten rock magma rose into the continental crust to form the granitic batholiths of the Sierra Nevada and Klamath Mountains. Evidently these began as basalt of the sea floor began to melt beneath the continent. As the new magmas rose, they in turn melted volumes of the old scraped-off sedimentary rocks of the continental margin to form molten granite magmas. The magmas also fed a chain of volcanoes similar to the modern Cascades that have since eroded away leaving their deepest roots exposed as batholiths. Painstaking studies of sandstones in the Great Valley Sequence have shown that some of them contain rock fragments erupted from these same volcanoes.

Long after the last tortured slice of Franciscan rock had been mashed into the Coast Range, probably sometime about 40 or 50 million years ago, the San Andreas fault system began to move. By this time the continental crust had solidly coupled to the rock beneath so that a moving segment of the ocean floor is now rafting a large slice of the Coast Range steadily northward at a rate of about 2 inches each year. This is an entirely different process from the one that operated during formation of the Coast Range when moving sea floor slid smoothly beneath the continent, scraping its blanket of sediments off against the edge as it disappeared.

Californians have worried about the San Andreas fault ever since it brutally devastated San Francisco in 1906. Frequent earthquakes here and there along its length remind everyone to keep worrying, the big fault is still there and still active, waiting for the inevitable day when it will strike another vicious blow. Meanwhile, geologists are trying to understand its movements hoping to learn how to predict and perhaps even to manage the earthquakes it generates.

Determining the direction and extent of movement along a fault is done by studying the rocks on either side to establish how far they moved to reach their present positions. In principle, a scissors and paste approach is used; if the map were cut along the line of the fault, how far would it be necessary to slide the pieces to match the rocks on opposite sides?

The large slice of granite west of the San Andreas fault in what geologists call the "Salinian Block" seems hopelessly out of place in the geologic picture of the Coast Range. If the state geologic map is cut along the line of the San Andreas fault and the western piece slid southward, opposite the direction it is moving, the granites of the northern Coast Range match those in the southern end of the Sierra Nevada

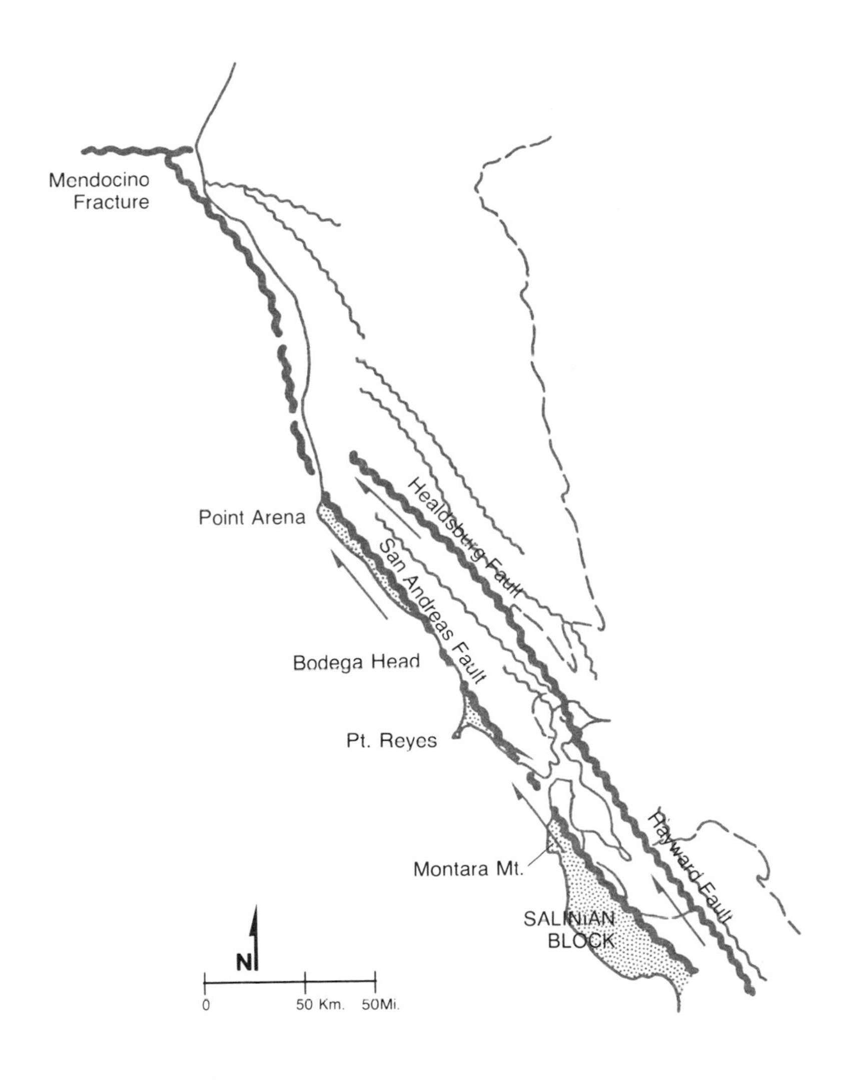

The "Salinian block" and several slices of the Coast Range are moving northward by movement along the San Andreas and related faults.

where they are amputated by the fault east of Los Angeles. That requires a displacement of about 350 miles. So it appears that the granites west of the San Andreas along the coast of northern California, from Montara Mountain to Bodega Head, were originally part of the southern Sierra Nevada. Each year they move approximately two inches farther north.

Northward movement of the "Salinian Block" is actually much too complex to be understood solely by displacements along the San Andreas fault. Jogs and bends in the line of the fault make it impossible to slide opposite sides of the map smoothly past each other without opening large gaps in some places and causing head on collisions in others. Evidently movement has involved a whole system of faults all sliding more or less in concert as they rip the soft edges of the two slabs of continental crust into a series of large slices approximately parallel to the main fault. Some of these branch faults are well known and precisely mapped but many others are not. Years of geologic work will be needed to unravel some of them and finally assemble all the stray pieces of the San Andreas puzzle of faults.

The last mangled slice of sea floor sediment and the last shattered masses of serpentinite were added to the Coast Range sometime toward the end of the Cretaceous Period, perhaps 80 million years ago. Subsequent additions have been local deposits of sandstone and mudstone supplemented by some patches of volcanic rocks.

Parts of the Coast Range have been shallowly submerged, while other parts formed islands and peninsulas, during much of the time since the rocks were squashed onto the edge of the continent. San Francisco Bay is a part of the Coast Range that is now submerged beneath sea water, just as other areas have been at times in the past. Sediments containing fossils of animals that live in sea water accumulated in the submerged areas, as they are now accumulating in the Bay, and remain as souvenirs of past inundations. In places they form deposits thousands of feet thick.

Much of the submergence happened during the last 15

million years — the period in which the San Andreas and related faults have been moving the "Salinian Block" northward along the coast and slicing much of the range into parallel ridges and valleys. Some of these slices sank temporarily parallel to the San Andreas fault, to form long lagoons, in which sediments accumulated. Many of the former lagoons are obvious today, even though not submerged, as flat-floored valleys elongated in a north-south direction; the Cotati Valley at Santa Rosa is a good example. Most of these valleys have also accumulated stream and hillslope sediments during times when they were above sea level.

Another event of the last 10 million years has been eruption of large quantities of light-colored volcanic rock in a sizeable area of the Coast Range centered around Sonoma and Napa Counties. These eruptions have coincided in time with movements on the San Andreas and related faults and it seems that there should be some connection but the relationship is not clearly understood.

Light-colored volcanic rocks are erupted mainly as ash because the molten lava is heavily charged with steam which blasts it violently out of the volcano as a cloud of small particles. These cool as they drift downwind and settle as thick deposits of ash, frequently burying forests which slowly become petrified wood. Several spectacular petrified forests have been found in the volcanic deposits of the Coast Range and the gravels of many streams contain pebbles of petrified wood.

Because they are very young, only a few million years old, many of the volcanic rocks of the Coast Range are still hot a few hundred feet below the surface. Rainwater percolating downward is heated and returns to the surface as hot springs and steam. One of these hot springs areas, the Geysers east of Cloverdale, was developed back in the 1920's into an electrical generating facility in which turbines are driven by natural steam produced from wells. Development has expanded in recent years and the plant now generates nearly half of San Francisco's electricity with heat reserves estimated to be sufficient for several centuries at least. No doubt future years will see much more widespread development of such facilities in other areas of California. There are, by the way, no geysers at the Geysers; the name is misleading.

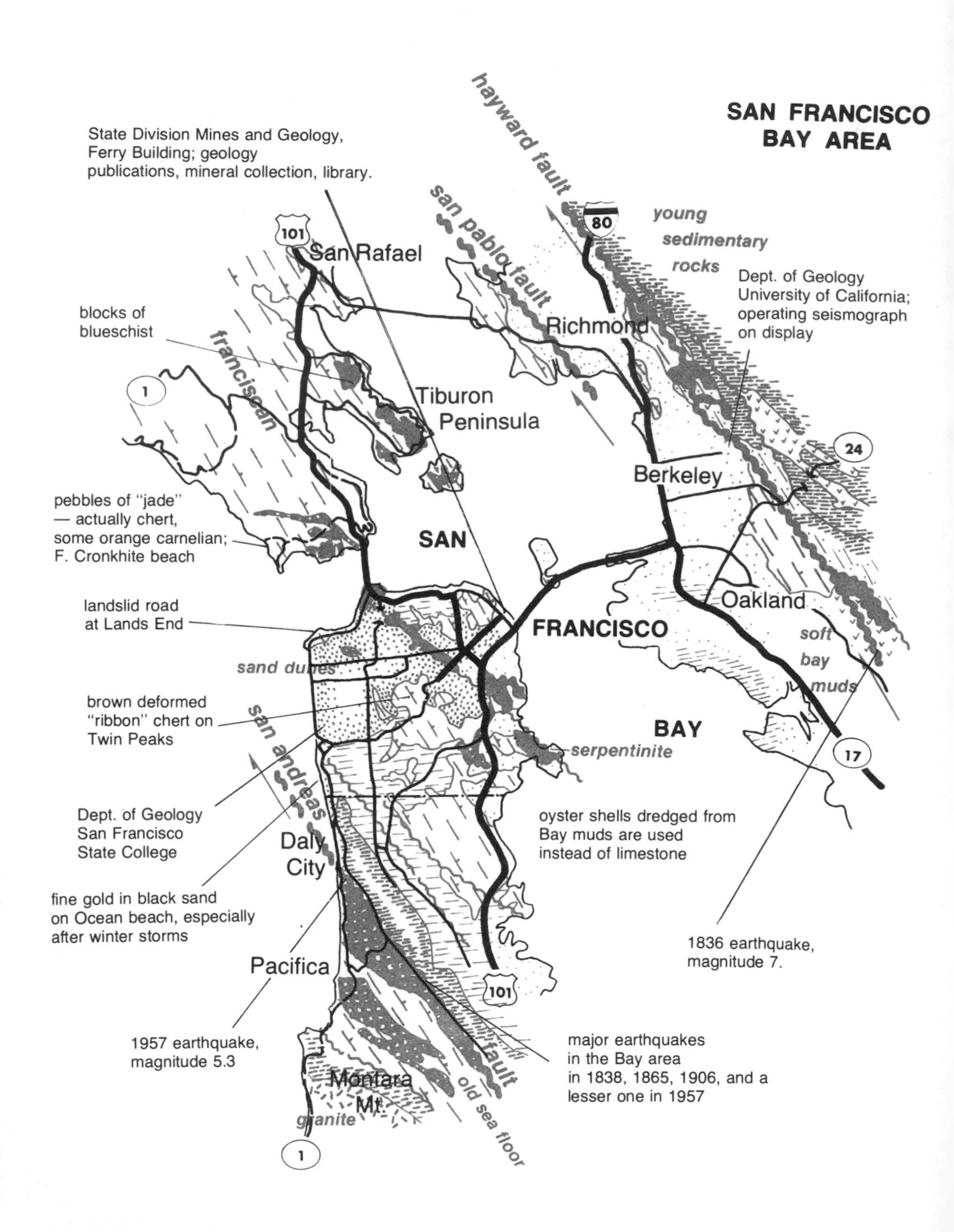

san francisco bay area

San Francisco Bay, one of the loveliest harbors in the world, is actually the drowned mouth of the Sacramento River. It was flooded when sea level rose 300 feet as the great continental ice sheets melted at the end of the last ice age about 10,000 years ago. Ever since then the river has been dumping its load of mud into bay, slowly but patiently filling it. Much of the flat land around the inland margins of the bay is sediment deposited within the last few thousand years. Eventually, thousands of years from now, the river will completely fill the bay with sediment, and grass will grow where ships now pass and the fog will drift over vast expanses of muddy swampland.

The Sacramento River carries drainage from most of the western slope of the Sierra Nevada across the Great Valley and through the Coast Range to the ocean. It is the only river that cuts right through the Coast Range and the only way it can possibly have managed to create such a course is to have found a way through while the mountains now in its path were temporarily below sea level. As the range slowly emerged, the river was able to maintain its course by eroding through the rocks as fast as they rose. Some of the sedimentary rocks the river cuts through in the Bay Area were deposited under sea water only a few million years ago so it seems that the Sacramento River must have established its course through the Coast Range sometime since then.

San Francisco Bay is very large in area and a little more than 300 feet deep in places so an enormous volume of water must flow in and out through the Golden Gate every time the tide changes. Scouring tidal currents keep the opening to the harbor clear of mud and complicate navigation through the Golden Gate. That passage was really tricky in the days of sailing ships which often had to wait days for a favorable turn of wind and tide to make the attempt.

Early settlers in San Francisco found much of the lower and seaward parts of the present city occupied by a spectacular tract of shifting sand dunes which they immediately began to cover with paving and buildings. Now most of the dunes are completely obscured except for a few lonely survivors in small areas in the general vicinity of the zoo. Here and there in Golden Gate Park the outlines of an old sand dune now covered by grass and trees is still vaguely visible to a sympathetic observer furnished with a knowing eye. Otherwise they are all gone and all that remains is sandy backyard gardens and occasional suspicious looking sandy humps.

All that sand must have been brought to the coast by the Sacramento River and then swept by the waves down the beaches extending southward from the Golden Gate. Sea breezes whipping across the drying upper beach during times of low tide carried the sand inland and heaped it into a choppy sea of shifting dunes. Of course all this must have happened a long time ago, before the mouth of the river was drowned by the flooding waters of the sea rising as the glaciers of the last ice age melted. Now the river dumps its sand into the inland reaches of the bay and the waves must depend for their supply upon the grains they can batter loose from the rocks along the coast or sift from the muds flushed from the bay in an outgoing tide.

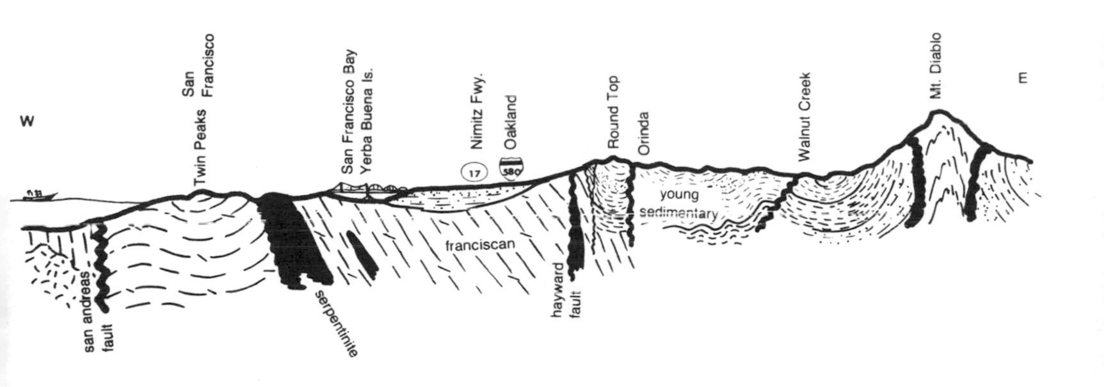

Section across the Bay Area showing how the Coast Range is sliced by faults. Mt. Diablo is a plug of Franciscan rock that pushed up within an enclosing shell of slippery serpentinite.

Bedrock in the Bay Area consists of long slices of the Coast Range slid into their present positions by millions of years of slow movement along the San Andreas fault and several of its branches which include the San Pablo and Hayward faults. These movements still continue and any of the numerous faults in the Bay Area is potentially the source of future earthquakes.

Rocks seaward of the San Andreas fault zone are granite, covered in many places by deposits of mud laid down beneath seawater during the last several million years. Montara Mountain, south of San Francisco, is a large knob of bare granite. These Coast Range granites are geologically misplaced; they belong to the southern Sierra Nevada before movement along the San Andreas fault transported them northward.

Except for the young deposits of unconsolidated sediment around the margins of San Francisco Bay and in the floor of the Santa Clara Valley, the outcrops between the San Andreas and Hayward fault zones expose Franciscan rocks typical of the northern California Coast Ranges. These underlie the hills of San Francisco and Marin County. They consist of a mixture of dark-colored muddy sediments, red, green and brown cherts and

Thinly-bedded Franciscan chert exposed in this Bay Area outcrop was once soft mud on the Pacific Ocean floor.

Fossil seashells in sandstone.

lava flows of black basalt — all material that was first laid down on the floor of the Pacific Ocean and then scraped into a trench at the ocean margin about 100 million years ago. Long zones of dark green serpentinite from deep in the earth's mantle, mark the lines of smaller faults within this slice of the Coast Range.

There are Franciscan rocks east of the Hayward fault zone too, but they are mostly buried under thick deposits of muddy sediments laid down beneath sea water while this area was temporarily submerged long after the Coast Range had formed. These weak, younger rocks, consisting mostly of thinly-layered mud and sand, weather and erode to form softly-rounded hills subject to constant landsliding. As urban development extends into these hills during years to come, landsliding will become a much larger problem unless stringent zoning and building code regulations are adopted and enforced.

Subsidence due to compaction of the soft muds beneath is causing many buildings in the Misson district of San Francisco to sink or tilt.

Serpentinite broken into shiny little fragments as the solid mass squeezed upward through Franciscan rocks.

Like many hilly cities, San Francisco has serious problems with landslides. Steep hillslopes and urban developers are natural enemies everywhere but things are especially bad in the Bay Area where deep soils developed on fractured bedrock make many of the slopes extremely unstable.

Cuts and fills to make level areas for building are almost unavoidable if an area is to be developed. Cuts make hills unstable by pulling support from under the slope above and fills make them unstable by adding an extra load to the slope below. Most hillsides can absorb a limited amount of this sort of abuse without sliding but many developments compound the damage by permitting excess water to soak into the ground in one way or another. Overflowing septic tanks, leaking swimming pools, lawn watering systems and inadequate storm drains all add water to the soil where it accumulates year after year, weakening the ground and adding to its weight. Eventually the slope takes its revenge by turning into a slumgullion of watery mud and rocks that slops down the hill converting everything in its path, roads, apartments, and houses into a disgusting mess of wreckage. Almost invariably the destroyed property contributed to creation of the slide by weakening the slope through cuts and fills and years of interfering with the natural drainage.

Old road near Lands End sliding piecemeal down the hill.

At any particular time there are dozens of slopes in the Bay area at various stages of the long process of becoming landslides, many of them near the terminal stage. All that is needed to get some action is one final insult — an unusually heavy run of spring rains, a light earthquake, or one last excavation job. Landslides are likely to follow in the path of almost any other kind of disturbance. The solution to the problem is to minimize cutting and filling on steep hillslopes and make every effort to keep them from absorbing excessive amounts of water. Some slopes are so delicately poised by nature that any development at all threatens them. These should remain forever untouched.

Drains stabilize road cuts in weak ground by keeping the soil dry.

The most fascinating geologic question to most people in the Bay area concerns the next big earthquake. Will there be another like that in 1906 and, if so, when will it come.

We know enough now about big faults and earthquakes to be sure that there will be more violent movements in the Bay area just as surely as one day follows another. The recent long years of quiet are not reassuring.

Recent investigations by the U.S. Geological Surgey suggest that the San Andreas fault is likely to produce a devastating earthquake in the Bay area at intervals of every 100 to 1000 years. The Hayward fault seems likely to release somewhat less devastating shocks considerably more frequently — every 10 to 100 years. However, these estimates are not the same as actual earthquake predictions; they are merely efforts to average the frequency of major shocks over a long period of time.

An easy way to visualize movement on a fault is to imagine two thick slabs of soft foam rubber sliding past each other with their cut edges touching. They move hitching along in sharp jerks as the edges of the slabs catch and stick together until the rubber stretches enough to snap them free. Faults move in the same way. Rocks on opposite sides catch, accumulate strain for a while by bending as though they were a giant rubber slab, and then pull free and snap past each other in a sudden jerk releasing the accumulated strain energy all at once as an earthquake. The trick in predicting earthquakes is to spot the places along a fault where opposite sides have caught, observe how fast and how far the rocks are bending, and try to anticipate the moment when they will finally come unstuck and release the earthquake. If a way could be found to free the rocks on opposite sides of the fault, or prevent them from sticking at all, then earthquakes could be managed or prevented.

As long as an active fault is moving freely the opposite sides glide smoothly past each other releasing frequent small earthquakes. Streets, sidewalks and walls built across the fault move steadily as the offset of opposite sides increases at a constant rate. As long as this is happening, there is little danger of a major earthquake because the rocks on opposite sides of the fault are not

Curb and sidewalk offset by slow creeping movement of the San Andreas fault.

R. L. Badley photo

accumulating strain by bending. Things become ominous when the steady movement and frequent small earthquakes cease because the opposite sides of the fault have stuck. Every day that passes without an earthquake in the Bay area adds a little more foreboding to the prospect of the one that will inevitably come someday.

Bending of the rocks near a stuck section of fault can be observed either by regularly resurveying lines across the fault or by directly measuring distortion of the rocks with delicate instruments called strain gauges. Such observations show how far the rocks have bent and how fast they are bending farther. From this it is possible to predict how far the fault will move and how violent the earthquake will be if it snaps loose today and how much farther it will move and how much more violent the earthquake will be at any future time.

Forecasting the day of an earthquake is still impossible because there is no way to know how firmly a fault is stuck and how much stretching will be needed to snap it loose. Another approach to the problem is to discover whether faults give some kind of advance signal that they are about to move — perhaps some kind of distinctive ground noise. Elaborate networks of seismograph stations in California keep an expensive ear to the ground listening for little creaks and groans in the hope that a way may be found to use them for predicting earthquakes. Perhaps seismograph observatories will someday be able to issue earthquake advisories as though they were weather forcasts but now we must still wait and expect them when they happen.

The prospect of a reliable method of earthquake prediction within the next few years conjures discouraging images of hordes of panicky people fleeing to the countryside leaving their homes and businesses to looters. Long range predictions would tempt unscrupulous operators to create economic chaos by plundering the real estate market. Inaccurate earthquake forecasts, and the first attempts are likely to be wrong, would cause needless consternation and skeptical indifference to future predictions.

No one has ever been killed or seriously injured by the jostling received from an earthquake. The dangers are from falling debris, landslides triggered by the quake, and fires and epidemics that may spread afterwards because the water supply is disrupted and contaminated. These are all manageable hazards.

Good construction is the best protection. Well-built wood or steel frame buildings and good reinforced concrete work are most un-

People live dangerously in these houses built on the San Andreas fault zone in the South Bay Area.

likely to come tumbling down during an earthquake even though they may shed some plaster and broken glass. Buildings with load-bearing masonry walls are likely to fall but very few of these exist in coastal areas of California, most of the old ones were shaken down in 1906 and most city building codes forbid them. The best earthquake protection for people in or around buildings is to crawl underneath something to shelter from falling glass or plaster. Immediately after an earthquake, and they only last a few minutes, it is a good idea to draw a supply of drinking water and prevent fires from starting or spreading.

Areas underlain by loose ground always shake more violently during an earthquake than those on solid bedrock for the same reasons that a gelatin salad quivers more than hard ice cream. Any low, flat place in the Bay Areais almost certain to be underlain by unconsolidated sediments that will shake like jelly. The Marina district is a good example of an area that will probably be devastated even though the buildings are sound because it is built on weak ground near the fault zone — it is an area that should have been made into a park. The nearby hills are eroded in solid bedrock and will shake much less.

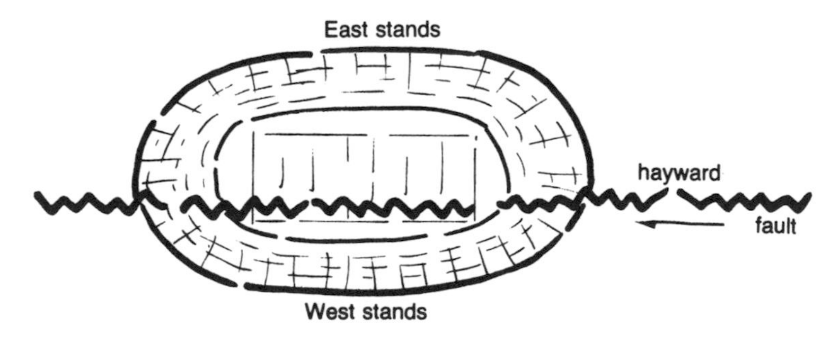

The Berkeley Stadium will be ripped apart by the next movement of the Hayward fault.

Neither is there much hope for buildings in the fault zone itself. The problem there is not just a matter of shaking but of having the ground torn as though it were a rag — no foundation can withstand that kind of treatment. Active fault zones are good places to build long, skinny parks and much of the San Andreas fault zone in the Bay area is developed that way. But parts of it are covered by residential and commercial developments which are doomed to certain destruction. The Hayward fault zone, which has also generated disastrous earthquakes within the last century, passes through many miles of town development in the East Bay area and may someday do terrible damage.

There is not the remotest glimmer of hope that any works of man will ever prevent a fault from moving. Nothing we can do will stop the restless movements of the continent as it rides the shifting currents of the earth's interior. No conceivable engineering structure can pin the pieces of a continent together any more than a real estate developer can eliminate a fault by landscaping a hillside.

Nevertheless, methods may someday be found to manage earthquakes by encouraging a stuck fault to slip early and release a small earthquake instead of waiting longer to build up to a major catastrophe.

An unintended experiment in earthquake management was accidentally performed a few years ago by the U.S. Army when it pumped surplus poisonous liquids down deep wells at the Rocky Mountain Arsenal near Denver. Several moderate earthquakes followed because the liquids lubricated a stuck fault, permitting it to move. Of course the fault would have moved eventually anyway, probably causing a more violent earthquake sometime in the indefinite future. Many geologists have imagined doing the same thing deliberately, releasing a stuck section of fault by lubricating it with water pumped into wells — water should do as well as surplus nerve gas. It seems likely that this would work but most unlikely that any agency would be willing to assume responsibility for unleashing the next San Francisco earthquake by trying out the idea. A better alternative would be to watch dangerous faults very closely and attempt to lubricate them as soon as they stick, before the rocks have bent far enough to be dangerous.

1

california 1 — the coast road

san francisco — leggett

The Coast Road winds along the seashore, poking into inlets and venturing onto bold headlands along the flank of the Coast Range where it slopes down into the ocean. It is a slow drive best suited for people disposed to stop for a walk on the beach or to clamber over sand dunes. Many of the rocks and sights along the coast are fascinating, as well as beautiful, and some of the pebbles are pretty enough to take home.

Most of the rocks exposed along the Coast Road are Franciscan sediments, mostly dark, muddy sandstones laid down on the floor of the Pacific Ocean about 100 million years ago and then sheared and folded as they were rudely stuffed into a trench at the continental margin by moving seafloor descending into the mantle. Occasional basalt lava flows interlayered with the sediments, and masses of green serpentinite that squeezed through them while they were being jammed into the trench, complete the inventory of major rock types exposed along the Coast Road. Some of the serpentinites contain chunks of blueschist, green eclogite, and jade, so pebbles of these attractive and resistant rocks are abundant in some of the streams and along many beaches.

Between San Francisco and Point Arena the road very closely follows the trace of the San Andreas fault which also influences long straight reaches of the coast. North of Point Arena the fault passes out to sea for the last time and the coast line, free of its sternly guiding influence, becomes much more sinuous and irregular. The trip is geologically fascinating not only because of the fault and the excellent exposures of bedrock along the sea cliffs but also because of the opportunity to watch the wind and waves actively doing the work of shaping the shoreline.

SAN FRANCISCO — BODEGA BAY
(63 miles)

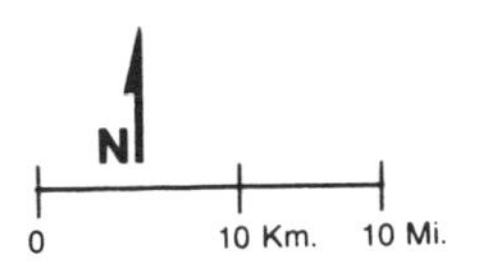

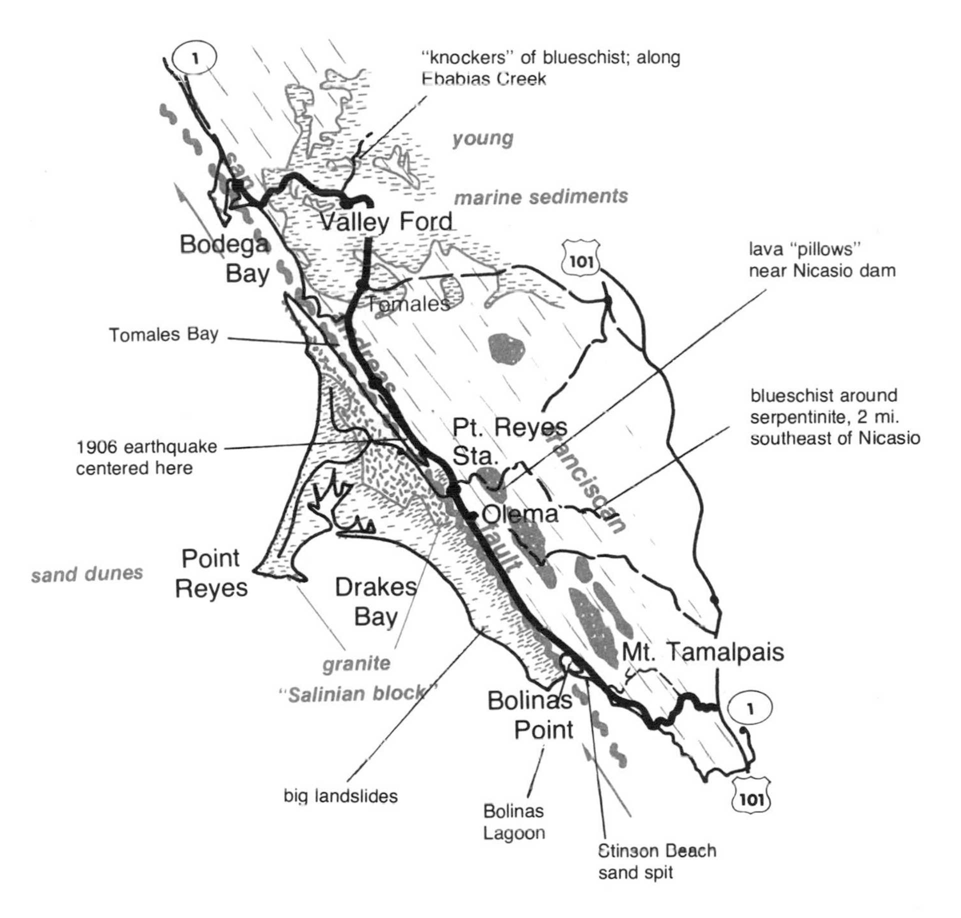

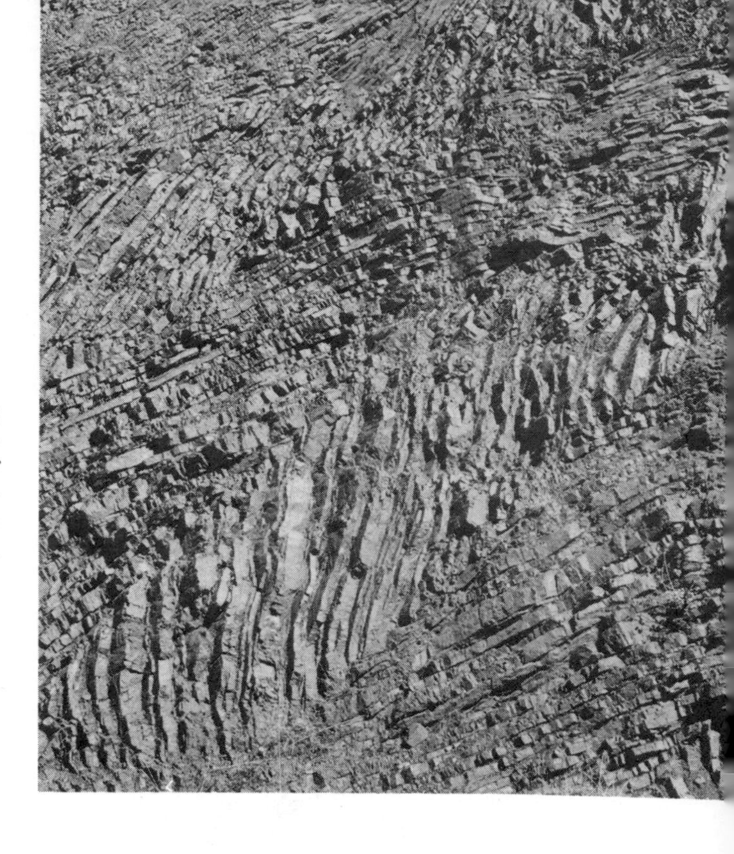

These crumpled ribbon cherts are exposed just west of the north abutment of the Golden Gate Bridge.

SAN FRANCISCO — BODEGA BAY

The long stretch of coast between San Francisco and Bodega Bay is the best area in northern California to observe the San Andreas fault. Most of the small details of surface displacement that appeared during the earthquake of 1906 have since disappeared but the major effects of large-scale faulting are perfectly visible and easily understandable.

Immediately north of the Golden Gate Bridge the Coast Road winds along the seaward margin of the Marin Peninsula, heading for its first rendezvous with the San Andreas fault at Stinson Beach. There are numerous outcrops of dark Franciscan rocks, mostly muddy sandstones and a few small exposures of thinly-bedded ribbon chert. Many small exposures of serpentinite are easily recognizeable by their dark color, various shades of dusky green, and criss-crossing pattern of fractures with shiny surfaces that glint in the sun.

Serpentinite on Mt. Tamalpais.

North of Stinson Beach the Coast Road follows the east side of Bolinas Lagoon, a drowned valley developed along the San Andreas fault zone. The road is built just east of the fault so all the outcrops near it are Franciscan rock, mostly muddy sandstones laced by closely-spaced fractures — eloquent testimony to the cruel grinding and crushing these rocks have suffered during the movement of the nearby fault. Rocks actually within the fault zone are broken and crushed almost beyond recognition, making them easy prey to the processes of weathering and erosion that have reduced the fault zone to a valley along most of its length. Hills on the far side of Bolinas Lagoon are underlain by granite; they are part of the "Salinian block" that has moved about 350 miles northward.

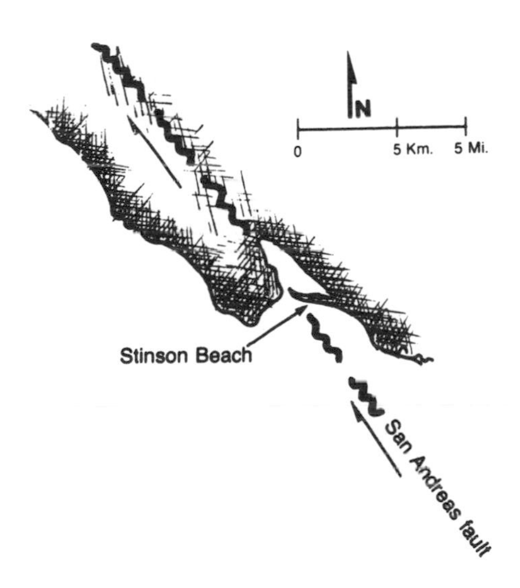

Stinson Beach is an example of a community that should have been located someplace else. It is built on a sandspit, a very unstable foundation material, almost directly over the San Andreas fault which passes under the western edge of the community and northwards up Bolinas Lagoon. It is difficult to imagine a combination of circumstances that could be more threatening if the fault were to move — as it certainly will.

1

At the north end of Bolinas Lagoon the floor of the valley rises above sea level and continues northward following a perfectly straight course. The road is close to the eastern margin of the fault zone. Surface movements of about 20 feet, as great as those measured anywhere, occurred along most of this valley during the 1906 earthquake. Roads, fences and lines of trees were offset and many buildings destroyed as the "Salinian block" abruptly jerked a bit farther north. One of the more macabre incidents of that earthquake happened on a ranch about two miles south of Olema where a cow fell into a fissure that briefly opened as the ground moved and was buried, except for her tail, when it slammed shut again. Despite the sad fate of the poor cow, which was well advertised, such calamities are actually quite rare during earthquakes and the danger of being swallowed up in the ground is hardly worth worrying about.

A short distance north of Point Reyes Station the fault valley drops below sea level again to become Tomales Bay. The road continues northward along the eastern side of the fault zone almost to the mouth of the bay before it jogs inland, away from the fault, to pass through the communities of Tomales and Bodega. It returns to the coast and to the San Andreas fault at Bodega Bay.

Fractured sandstone.

Bodega Head, like the Point Reyes Peninsula, is a slab of granite, another scrap of the "Salinian block" moved north along the San Andreas fault. Granite outcrops at the seaward tip of Bodega Head are connected to the mainland by a broad swath of wave- and wind-deposited sand covering the actual fault zone. This is the only place between the head of Tomales Bay and Fort Ross where the fault actually passes beneath dry land — it is just offshore along most of this part of the coast. Much of the community of Bodega Bay, like Stinson Beach, is built on loose sediment deposited directly on top of the San Andreas fault zone. It is another community clearly destined from the day the first house was built to be destroyed every time the northern part of the San Andreas fault releases a major earthquake.

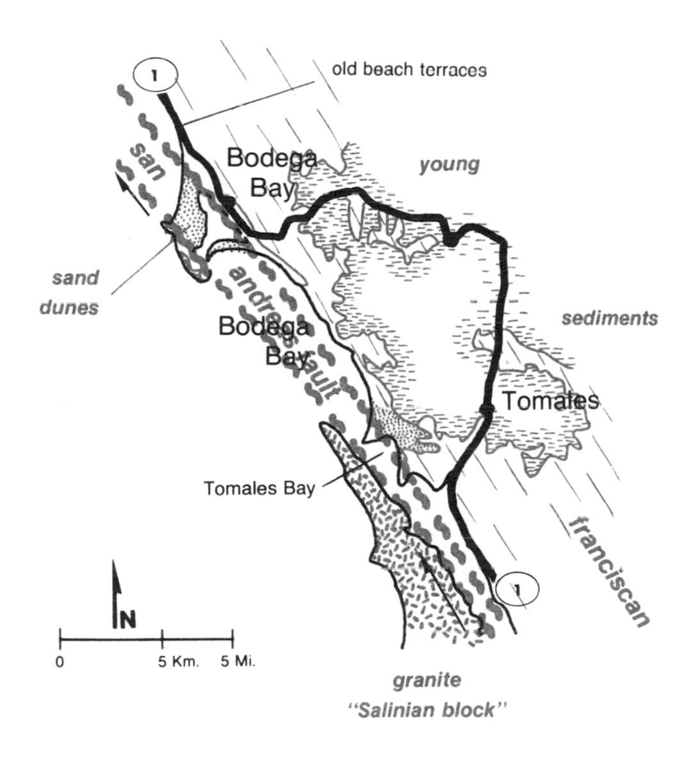

1

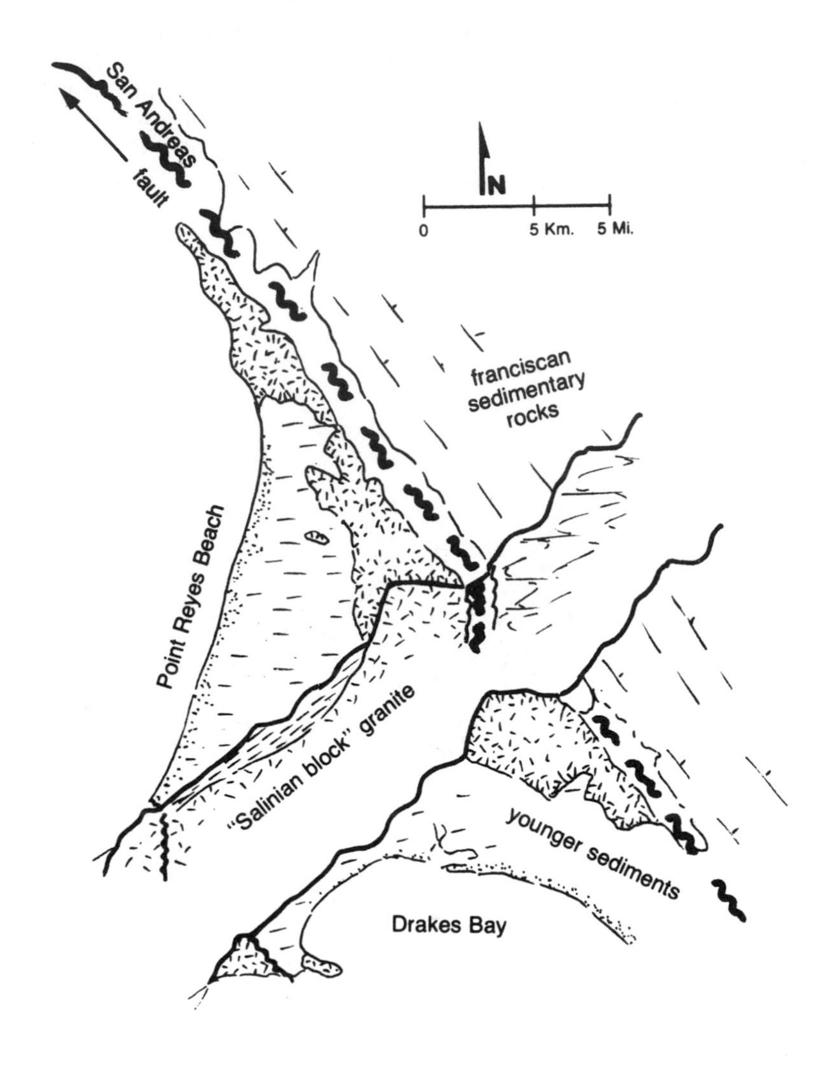

Younger sediments bury the granite except along the inner edge and outer tip of Point Reyes.

POINT REYES NATIONAL SEASHORE
— wind, surf and sand

The map of the Point Reyes peninsula looks like an angular afterthought dabbed at the last moment onto the coast of California. That is exactly what it is, a stray scrap of Sierra Nevada granite moved about 350 miles northward by displacement along the San Andreas fault. Small outcrops of weathered granite, well shrouded in vegetation, are visible beside the road and in the woods along the west side of Tomales Bay. Bold outcrops of granite form the towering sea cliffs that buttress the extreme tip of the peninsula around the Coast Guard station. In the remainder of the peninsula the granite is buried under deposits of sedimentary rock laid down beneath sea water between 10 and 20 million years ago. These are magnificently exposed in the long line of pale sea cliffs that face south across the sheltered waters of Drakes Bay.

Waves may travel across half an ocean or more from the place where they gather their energy from the winds at sea, to the coast where they finally spend it in doing the work of shaping the beach. They sense the shore before they reach it by dragging on the bottom as they get into water shallower than the distance from wave crest to wave crest. There the waves slow down and crowd closer together as their crests pile higher to form the curling breakers that crash themselves against the coast.

Because their headlong rush to the coast is slowed by drag on the bottom as they enter shallowing waters, waves reaching the shore at an angle, hold back near the beach and forge rapidly ahead where they remain in deeper waters offshore. So the waves of the open ocean change their shape to conform to the outline of the shore. They curl tightly around the tips of jutting headlands and stretch themselves out to extend along the shorelines of deep bays.

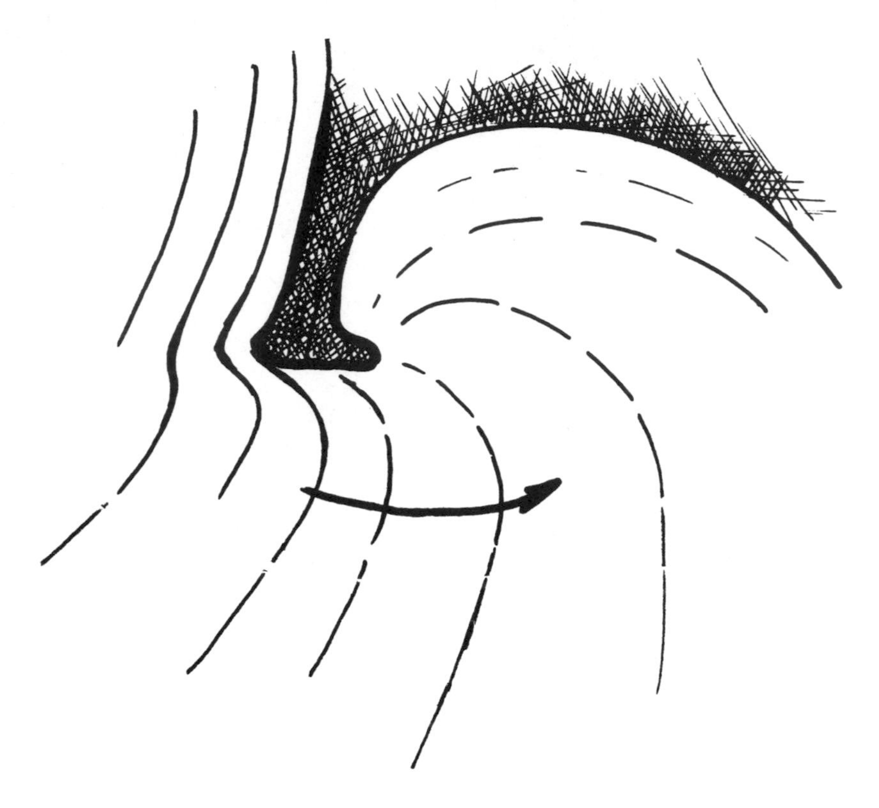

Point Reyes Beach, on the straight northwest side of the triangular peninsula, is constantly pounded by a heroic surf. The prevailing wind builds the waves into heaping rollers as it drives them across the thousands of miles of unhindered ocean between Siberia and California. Smaller waves that lap gently along the southfacing curve of shoreline along Drakes Bay are actually the same ones that pound directly onto the north side but stretched thin as they wrapped themselves around the entire peninsula to embrace this sheltered coast. Huge breakers that crash constantly against the rugged granite cliffs at the extreme tip of the peninsula are the same waves again, their energy focused on the cliff as they curl around it.

Waves crashing into a sea cliff drive pulses of compressed air into cracks in the rock popping it loose block by block and cutting a notch into the base of the cliff. When the notch gets deep enough, the entire face of the cliff collapses into the sea and the waves clear the rubble away as they start undercutting the new cliff face. Thus the waves cut a smooth shoreline and gently shelving beach into the mountainous coast of California. Jutting headlands are attacked from all sides at once as the waves curl around them upon approaching the shore. Sediment derived from their disintegration is swept into nearby bays as the originally rough shoreline is converted into a series of smoothly curving beaches.

Where sandy beaches receive the swash of the breakers, each incoming wave sweeps sand shoreward as it breaks and then washes it back toward the sea as it withdraws. If the waves happen to approach the coast at a slight angle, as they usually do, they also move sand along the beach. It is easy to watch this by following the progress of a seashell as each breaker washes it a few feet along the beach with each passage shoreward and back. Every grain of sand moves in the same way. So the waves may move the entire beach along the coast and the sand at one place today will be someplace down the beach tomorrow and replaced by new sand freshly arrived. This process must have provided Point Reyes Beach with most of its sand which consists of small red and green pebbles of chert obviously eroded from Franciscan rocks exposed along the coast to the north and on the opposite side of the San Andreas fault. There are no such rocks on the Point Reyes peninsula.

Waves wash sand high onto the beach when the tide is in and leave it there to dry in the sun when the tide is out. Sea breezes blowing across the beach at low tide pick up the sand and sweep it inland where they heap it into dunes high above the reach of the waves. Sand dunes are creatures of the beach just as much as of the desert. There are nice ones all along the north side of the Point Reyes peninsula, some of them actively moving and others stabilized by a cover of vegetation.

The very existence of sand dunes is one of those uncelebrated natural marvels that we blithely accept without pausing to wonder why the winds should pile sand into dunes instead of spreading it all over the countryside. Wind driven sand moves by bouncing and skittering along the ground surface, never rising more than a few inches into the air. It sweeps rapidly across hard surfaces and slowly across soft surfaces where the grains tend to stick instead of bouncing. Sand dunes have soft surfaces which trap grains of blowing sand, adding them to the dune and preventing their being blown all over. Much harder surfaces good for continued move-

Pebbles and damp sand make a hard surface on the beach.

ment of sand are common along the upper beach in the form of patches of damp sand, accumulations of seashells, or larger pebbles.

One of those deplorable spectacles common where people have dealt heedlessly with natural places is a squalid mess of old bottles or cans paving the bottom of a hole in a dune. Someone discarded the junk on top of the dune creating a hard place in its naturally soft surface. This helped transport of sand causing the wind to excavate the hole which will last as long as the hard objects are exposed but will be refilled by the wind if they are covered by a thin layer of soft sand. A soft surface is the very essence of a sand dune. Vegetation stabilizes sand dunes halting their onshore march but does not prevent them from trapping more sand and growing larger. It cannot be the roots of the plants that hold sand, as many people assume, because wind transport of sand happens above the surface where no roots exist. It is the leaves that stabilize the dune, breaking the force of the wind and absorbing impacts of incoming sand grains to prevent them from kicking other grains up into the wind.

Straight ripples rule the surface of a dune.

A fascinating struggle is constantly waged between wind and plants for possession of the dunes. Many of the dunes above Point Reyes Beach have been cloaked in greenery for years and have become shapeless heaps of sand. Others are clear of plants and march freely inland, rakishly sculptured by the driving wind. Most of the wind-blown sand is in irregular heaps and hollows piled around plants and blown out from between them.

Chunks of solid sandstone stand like monuments in the San Andreas fault zone near Tomales. Surrounding rock is crushed.

BODEGA BAY — POINT ARENA
(61 miles)

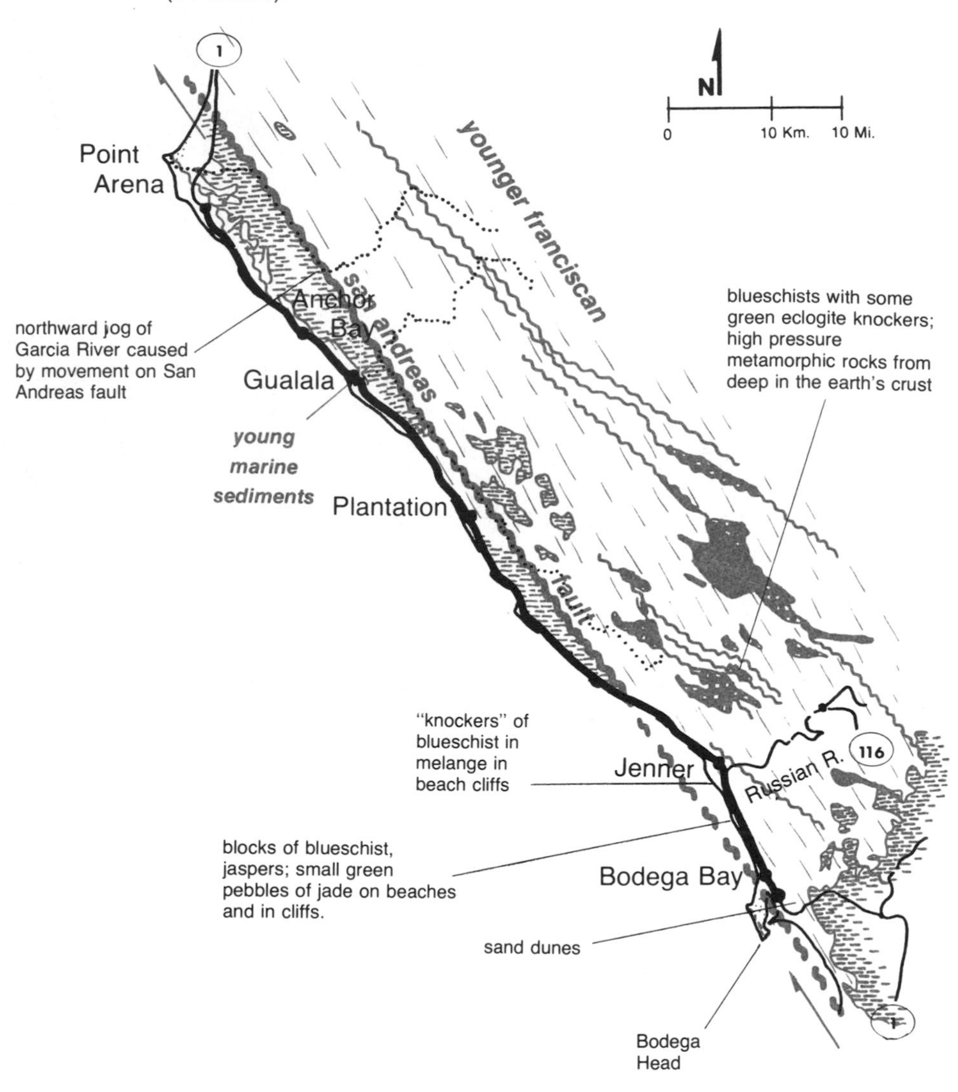

Solid blocks of sandstone swim in a matrix of crushed rock in this "melange" exposed in a sea cliff 6 miles north of Bodega Bay.

BODEGA BAY — POINT ARENA

Between Bodega Bay and Fort Ross the San Andreas fault is offshore and the coastline is entirely in Franciscan rocks. Serpentinites near Jenner are well known for the large chunks of unusually beautiful blueschist they contain — watch the roadcuts immediately south of town.

Just about two miles south of Fort Ross, the San Andreas fault comes ashore — exposures of terribly crushed and broken Franciscan rock in roadcuts near the mouth of Timber Gulch mark the place where it crosses the coastline and disappears into the wooded hills to the north. Extensive damage done in the Fort Ross area during the 1906 earthquake included splitting several large redwood trees that happened to be growing directly over the line of the fault in the woods north of town. Recent detailed study of these trees has shown that they had also been damaged by previous earthquakes and survived, preserving the record of their adventures in their patterns of annual rings.

1

Along most of the distance between Fort Ross and Point Arena, the San Andreas fault zone is expressed as a gentle valley following a perfectly straight course between 2 and 5 miles inland. The small community of Plantation is in the fault valley and the Gualala and Garcia Rivers both turn and follow it, flowing parallel to the coast for some miles before finally entering the ocean.

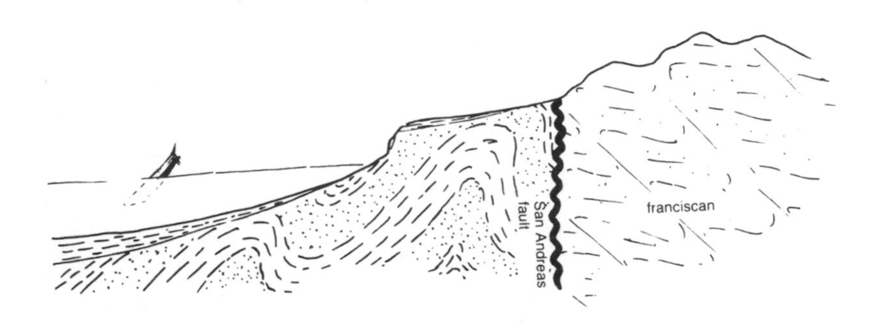

Section across the coast between Bodega and Point Arena. Rocks west of the fault are sediments deposited near Santa Barbara.

The Coast Road follows a route on the "Salinian block" west of the San Andreas fault along the entire distance between Fort Ross and Point Arena. Rocks exposed along the coast west of the fault, along the line of the road, are folded sedimentary layers deposited in the ocean along the margin of the continent between 70 and 20 million years ago. No granite is exposed in this part of the "Salinian block" and it seems likely that none exists even at depth — all available evidence suggests that these rocks rest directly on the sea floor. Studies of unusually distinctive dark pebbles contained in some of the layers exposed near Anchor Bay show that they were probably eroded from bodies of similar rock that form part of the Temblor Range east of Santa Barbara. Evidently this part of the "Salinian block" was in that vicinity when these rocks were deposited and has since moved about 350 miles north. Several other similarities across the San Andreas fault in other areas give about the same result.

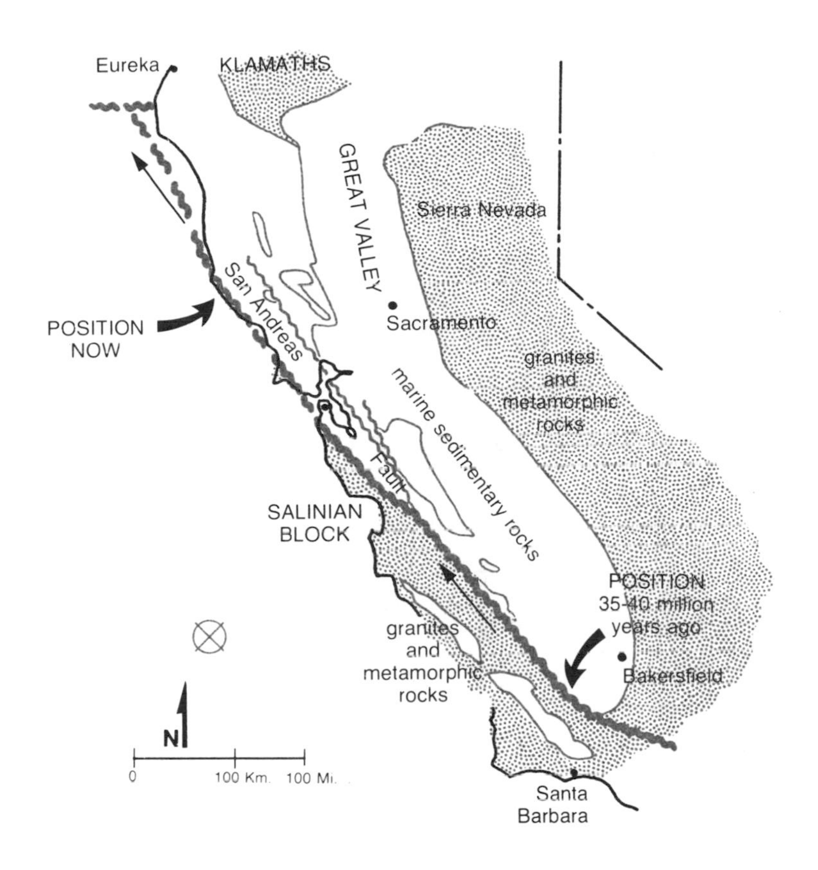

Rocks exposed along the Mendocino Coast began their careers near Santa Barbara.

Approximately 100,000 repetitions of the amount of displacement observed during the earthquake of 1906 would be required to move the "Salinian block" the distance it appears to have travelled. If there were one earthquake like that of 1906 every century or two for 10 or 20 million years, the job would be done. Much of the available evidence suggests that some such figures must describe what has happened.

The San Andreas fault passes out to sea for the last time at the mouth of Alder Creek about 12 miles north of Point Arena. The fault follows the valley immediately east of Manchester Beach State Park and from there northward closely parallels the coast for some miles, gradually getting farther out to sea as the coast trends away from it. Ultimately it meets the Mendocino fracture, some miles off Cape Mendocino and disappears westward in the floor of the Pacific Ocean.

1

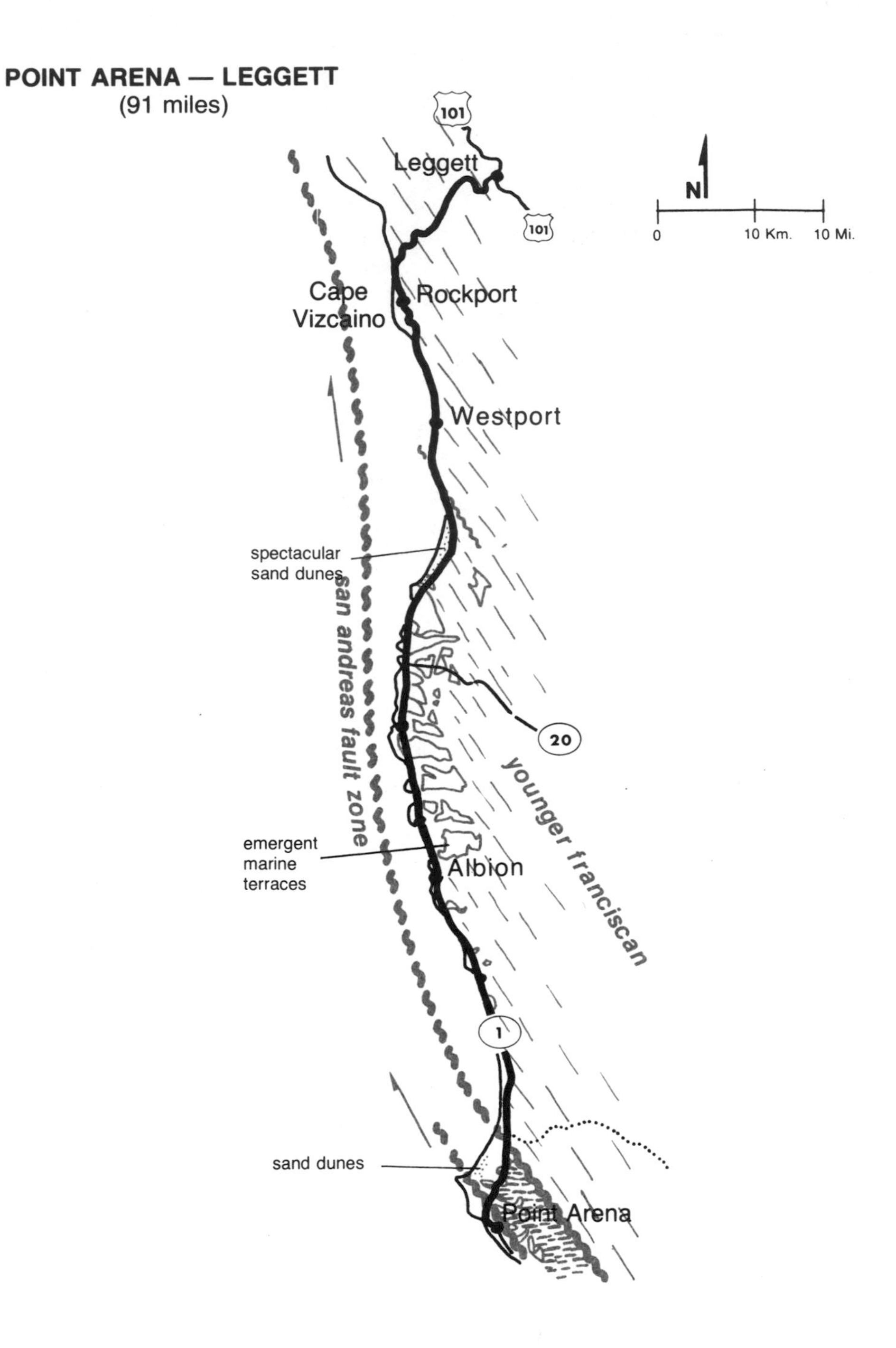

POINT ARENA — LEGGETT
(91 miles)
101
Leggett
101
N
0
10 Km.
10 Mi.
Cape Vizcaino
Rockport
Westport
spectacular sand dunes
san andreas fault zone
20
younger franciscan
emergent marine terraces
Albion
1
sand dunes
Point Arena

Mendocino is built on a marine terrace.

POINT ARENA — LEGGETT

Only Franciscan rocks are exposed along the coast north of Alder Creek where the San Andreas fault passes out to sea for the last time. Most of these are dark, muddy sandstones but some layers are lighter colored and locally, there are red and green cherts. Along many long stretches of this coast the steeply tilted Franciscan rocks are buried beneath gravelly younger beach sediments deposited fairly recently and still not consolidated into hard rock.

Emergent marine terraces fringe most of the California coast but are especially conspicuous along this stretch of road. Almost the entire length of the coast road is built on them, smooth land surfaces — usually farmed — that slope gently down toward the sea and end abruptly at the top of a recently eroded sea cliff. These gently sloping benches along the coast are old beaches planed to their present shape by wave action and then raised above sea level by uplift of the Coast Range.

Most coast lines have marine terraces along them and geologists invariably dispute whether they mean that the land has raised or

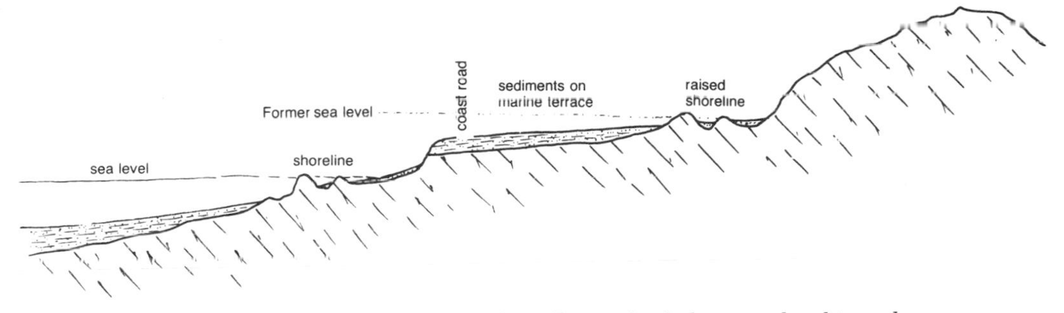

Section across the coast showing old sea floor raised above sea level to make a marine terrace.

Flat-topped remnants of a marine terrace being carved up by modern wave erosion on the beach at Fort Bragg.

sea level dropped. It is hard to know because the terrace looks the same in either case. Fluctuations of sea level must have helped form the terraces along the California coast but uplift of the land probably played a more important role. Sorting out these factors is surprisingly difficult.

Between Fort Bragg and Westport the Coast Road continues to follow the surface of an old wave-cut bench now raised above sea level and become a marine terrace. There are relatively few bedrock outcrops along the road but the sea cliffs below road level contain numerous excellent exposures of Franciscan rocks.

Immediately north of the community of Cleone, between 5 and 10 miles north of Fort Bragg, is one of the most spectacular tracts of sand dunes on the northern California coast. Several square miles of dunes cover the area between the road and the beach along an

Fractured sandstone sculptured by weathering – by the beach at Fort Bragg.

The wave-planed marine terrace slopes gently down to the modern sea cliff near Westport.

especially beautiful stretch of shoreline. A few dunes come almost to the road but it is necessary to stop for a walk to the beach to really enjoy them. It seems likely that they are here because the Tenmile River, which empties into the sea t at the northern edge of this dune field, dumps its load of sand on the coast at a place where the prevailing winds can blow it back onshore.

About five miles north of Westport the Coast Road leaves the seashore and turns inland towards Leggett. The last views north along the coast towards Cape Vizcaino reveal steep mountain slopes descending directly into the foaming surf, unbroken even by the slightest hint of an emergent wave-cut beach — no easy route for road construction. This seems to be an important change because the emergent marine terraces so conspicuous along most of the California coast do not look the same north of Cape Vizcaino. Perhaps this abrupt change in the nature of the coastline is related to the fact that the Mendocino fracture, a boundary between major slabs of the Pacific Ocean floor, meets the continent about 30 miles farther north at Cape Mendocino. Movements of the continental crust south of the Mendocino fracture appear to be raising the

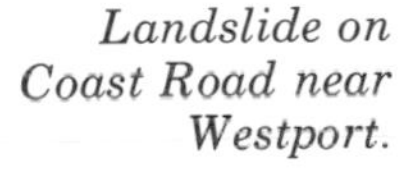

Landslide on Coast Road near Westport.

Coast Range, creating emergent marine terraces. The slab of the Pacific Ocean floor north of the Mendocino fracture appears to be descending into the mantle just offshore, pulling the Coast Range down a bit and preventing emergence of the shoreline.

Between the coast and Leggett the road winds tortuously along a corkscrew route through a steep Coast Range landscape corrugated into an endless succession of sharp ridges and deep ravines. Dense second growth forest sprouted since the original giant redwoods were cut, grows luxuriantly in deep soils developed under the influence of a warmly humid climate. The red iron oxide color of these soils, suggest that they are leached of fertilizer nutrients and too poor for any crop but trees. Slow downward soil creep and landslides are the erosional processes chiefly responsible for shaping these hills.

Steeply-tilted Franciscan sandstones exposed in a roadcut near Leggett.

u.s. 101 — the redwood highway

san francisco — oregon line — landslide country

The Redwood Highway winds through rolling, forested hills of the Coast Range all the way between San Francisco and the Oregon line. Almost all of the outcrops along the way are Franciscan rocks, the deep sea sediments that were stuffed into a marginal trench about 100 million years ago and later rose to become a range of low mountains fringing the continent. When it first rose above sea level, perhaps 70 million years ago, this range was an offshore peninsula and chain of islands separated from the mainland by an inland sea that has since filled brimful with mud to become the Great Valley.

Deep soils mantle most of the Coast Range, covering the bedrock and giving the hills their softly-rounded shape. Millions of years of weathering under the benign but rock-corroding influence of a warmly humid climate produced these soils while the protective umbrella of the dense forests sheltered them from the eroding impacts of raindrops. Their reddish colors, the characteristic shades of iron oxides, betray their poverty — millions of warm rains have washed away the soluble fertilizing nutrients, leaving little behind but the inert oxides of iron and aluminum, an inhospitable residue suitable for growing trees and very little else. Only the bottom lands floored by stream-deposited soils and some of the less thoroughly leached soils in the drier, eastern parts of the range are naturally fertile for agricultural crops.

Thick soils developed on deeply weathered bedrock are naturally likely to slide, especially if the bedrock is as weak and broken as the Franciscan sediments that underlie most of the Coast Range where landslides are so frequent that they seem to be one of the most important processes of erosion. Soil slumps off the hillslopes and into the streams which function as the conveyor belts of the landscape to haul it away to the sea. Bedrock freshly exposed by the landslide slowly weathers to make new soil which in turn will eventually move down the slope into the streams.

Most stretches of U.S. 101 pass numerous landslides, a few of which are perfectly natural, but many of which involve slopes disturbed by construction. Most slopes stand at angles gentle enough to safely hold their cover of soil without much danger of sliding. Construction projects often cause previously stable slopes to slide either by making them steeper or by soaking them with water.

Excess water makes soil sloppy and weak, greatly increasing the danger of sliding. In most areas the natural vegetation uses more water than crops or pasturage so cutting the forest and converting land to agriculture may diminish the amount of water drawn from the ground by plants thus causing the soil to become wetter and weaker. Any kind of interruption of the surface drainage may increase the water content of the soil and cause sliding.

Excavation on a hillside steepens the slope above by undercutting its base and the fill steepens the slope below by loading its top.

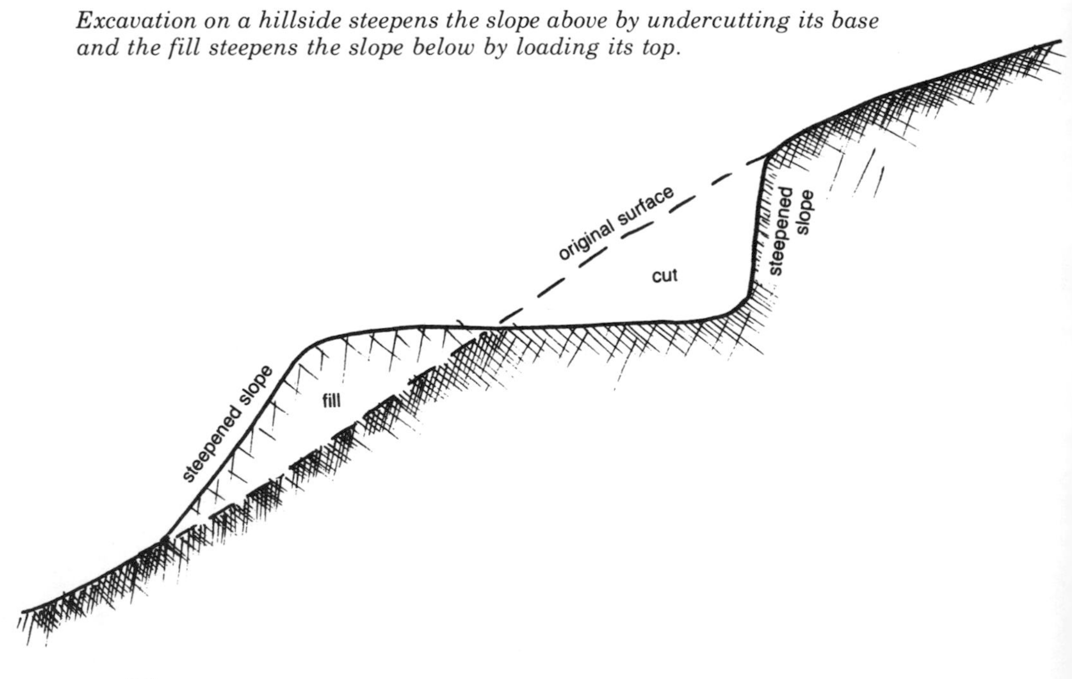

Landslides frequently damage highways in the Coast Range.

When a slope gets ready to slide, a curving fracture shaped like the bowl of a spoon develops within the soil. Cracks open in the ground at the upper end of this fracture and the slope bulges at the lower end. The slide mass above is unstable but it may not move for years. Or it may begin to move very slowly, perhaps a few inches per year. Some landslides move slowly for many years until a series of heavy rains or an earthquake loosens them up and dumps the whole mass down the slope. The slide mass may move as a relatively solid chunk of ground, but most of them contain a lot of water and are likely to slither down the slope as a loose pudding of sloppy mud.

Whether they move very slowly or all at once, landslides leave a curving scar, the fracture surface, on the hillside where they started and a hummocky dump of debris beneath it. Sets of curving fractures at the upper part of the slide mass and a wrinkled and lumpy look near the lower part betray a slope that is beginning to slump. Evidence that a slope is now sliding or has recently done so is usually obvious and should warn anyone not to attempt development. Nevertheless, millions of dollars worth of property are destroyed every year in the northern Coast Range by landslides that could have been avoided. In most cases the injury is not due to some unpredictable natural event but to human ignorance or stupidity.

Drains help prevent landslides by keeping the slope dry.

Landslide damage can be avoided by choosing stable hillslopes for construction and then taking care not to oversteepen them or saturate them with water. Cuts should be graded to gentle slopes and no large fills piled on steep hillslopes. Efforts to prevent infiltration of water and provide good drainage will minimize the danger of slides.

Piles of rock loaded onto the toe of a landslide will often stop it in its tracks.

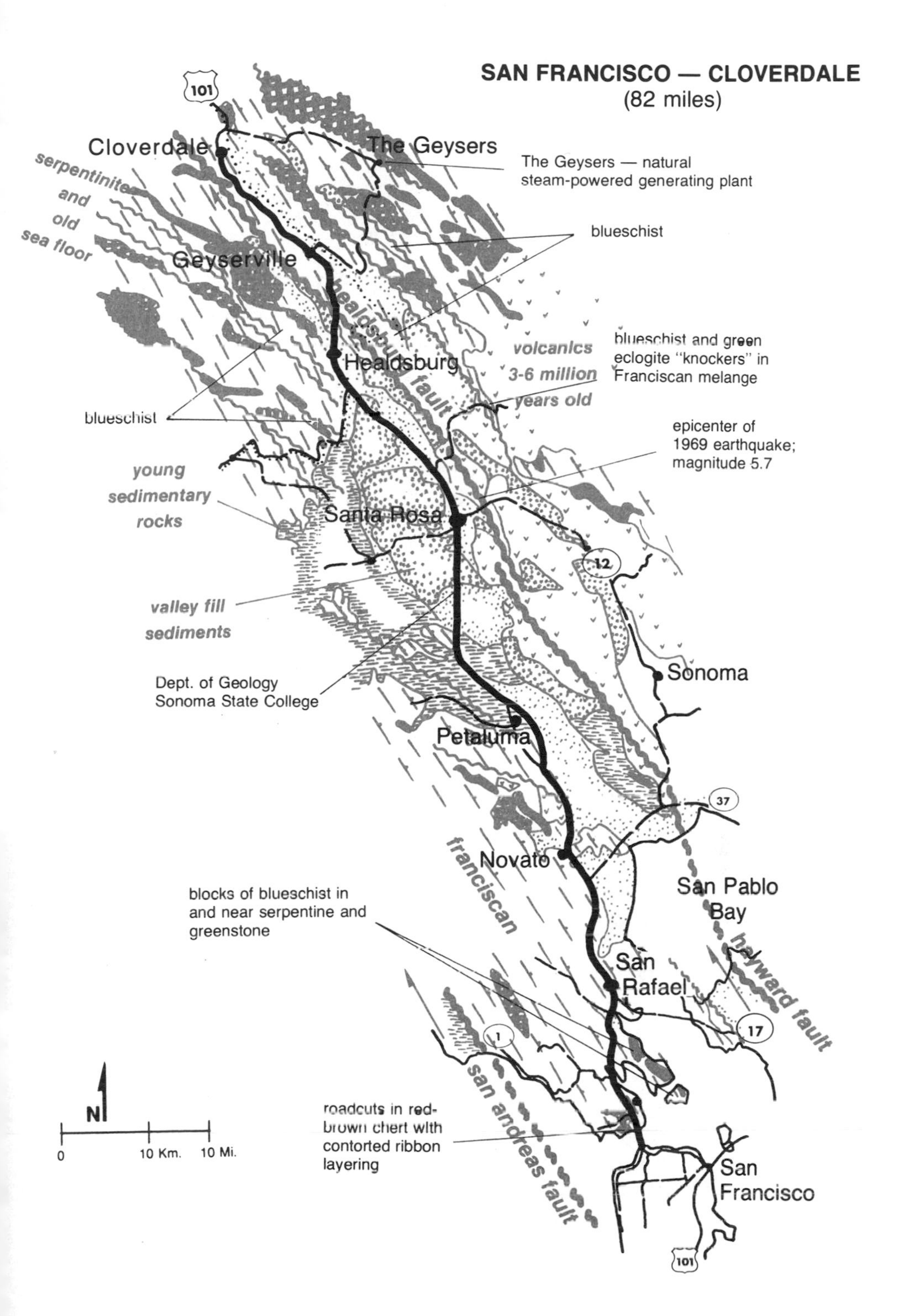
SAN FRANCISCO — CLOVERDALE
(82 miles)
101
Cloverdale
The Geysers
The Geysers — natural steam-powered generating plant
serpentinite and old sea floor
Geyserville
blueschist
Healdsburg
healdsburg fault
volcanics 3-6 million years old
blueschist and green eclogite "knockers" in Franciscan melange
blueschist
epicenter of 1969 earthquake; magnitude 5.7
young sedimentary rocks
Santa Rosa
12
valley fill sediments
Sonoma
Dept. of Geology Sonoma State College
Petaluma
37
Novato
franciscan
San Pablo Bay
blocks of blueschist in and near serpentine and greenstone
San Rafael
hayward fault
17
1
san andreas fault
N
0
10 Km.
10 Mi.
roadcuts in red-brown chert with contorted ribbon layering
San Francisco
101

SAN FRANCISCO — CLOVERDALE

Along the route between San Francisco and Cloverdale, about 80 miles, U.S. 101 passes through some of the most varied and interesting rocks in the northern California Coast Range. Unfortunately good bedrock outcrops are few so most of the geologic features must be visualized from the landscape.

Some outstanding exposures are beside the freeway immediately north of the Golden Gate bridge: large roadcuts between the Marin abutment and the Sausalito interchange. Usually it is impossible to pause for a good look at these without risking a traffic accident in the fog but Sunday afternoon bottlenecks north of the bridge sometimes provide the opportunity. Rocks in these roadcuts are a good sample of the Franciscan sediments that form most of the northern Coast Range. There are numerous layers of muddy sandstone, dark rocks drably lacking character in their appearance, and a few flows of basalt lava, their original flat black color now streaked with shades of dull green. Layers and occasional globs of red are chert, a very hard rock composed of the accumulated quartz skeletons of unimaginable billions of microscopic animals welded by recrystallization into a solid mass. The road west to Point Bonita from the north end of the bridge passes beautiful outcrops of chert. All of the sedimentary layers tilt steeply down to the east and show obvious signs of having been torn, sheared and crumpled — deformation that occurred as they were scraped off the surface of the descending sea floor and stuffed against the continental margin about 100 million years ago.

Large masses of serpentinite that include beautiful chunks of dark blue blueschist and dark green eclogite form the backbone of the Tiburon Peninsula and part of Angel Island. Collecting used to be wonderful on the Tiburon Peninsula but suburban development has already buried many of the best localites and others disappear beneath parking lots and apartment complexes with depressing regularity. Good specimens are easy to find on Angel Island but that is a park so collecting is not considered public-spirited.

A landslide hollowed the side of this hill in the northern Coast Range.

Highway 101 crosses Franciscan rocks, locally exposed in roadcuts, as far north as the Petaluma interchanges where it enters the south end of the large Cotati Valley which extends northward to Healdsburg. Because it is elongated parallel to the trend of the San Andreas fault system and at least partly bounded by faults belonging to that system, it seems likely that movement along these faults several million years ago first opened the low area in the Coast Range that we know today as the Cotati Valley. Events since then have changed it greatly and largely hidden the evidence of how it first formed.

Low hills forming the western rim of the Cotati Valley, west of Santa Rosa Creek, are eroded into beds of soft sedimentary rock—essentially undisturbed layers of mud that cover the intensely-deformed Franciscan rocks over an area of several hundred square miles extending almost to the coast. These muds contain abundant fossil remains of animals known to have lived in shallow seawater during the Pliocene Period, the span of time between about 15 and 3 million years ago. Evidently a shallow-water bay of the ocean

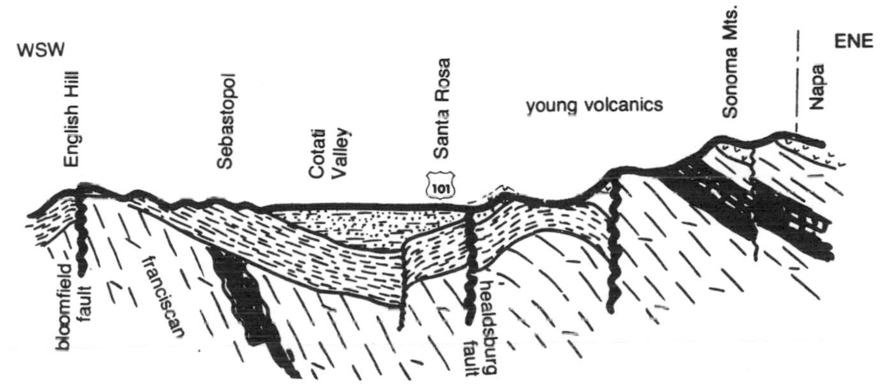

Section across the Cotati Valley Near Santa Rosa. Deep layers of younger sediments bury the Franciscan rocks in this area.

existed here. Higher hills that form the eastern rim of the valley, the Sonoma Mountains, are eroded into volcanic rocks, mostly deposits of light-colored volcanic ash erupted while the valley was a bay. In places these contain petrified trees, remains of lush subtropical forests overwhelmed and buried by volcanic eruptions. We can reconstruct a tempting memory of the Cotati Valley as it must have been during the Pliocene Period — a broad saltwater bay rimmed on its north and south sides by low hills eroded in Franciscan rocks and nestling beneath high volcanoes to the east — all the land cloaked in dense green forests.

After the bay had filled and drained and the volcanoes had become extinct, streams eroded broad valleys deep into the rocks that record their former presence. Later these valleys were partially filled by gravel washed in from the surrounding hills and now streams are cutting new valleys deep into those gravel deposits. It is these broad aprons of gravel washed into the stream valleys from nearby hills that form the basis for the rich and well-drained soils of the Cotati Valley. Without them the vineyards could not flourish.

North of Healdsburg the highway leaves the Cotati Valley and crosses into the much smaller Alexander Valley in which it follows the Russian River to Cloverdale. Although it is smaller and never contained a bay snuggled at the base of high volcanoes, the Alexander Valley also appears to be related in origin to movements on the San Andreas system of faults. Hills bordering it are eroded into Franciscan rocks.

The big geothermal power plants at "The Geysers" begun in 1955 are in Franciscan sandstone and basalt 8 miles northeast of Geyserville along the north side of Big Sulphur Creek. The "geysers" are actually hot and blowing steam springs, natural steam boilers, which have now been converted to steam wells used for generating electricity. The springs, discovered by a hunter in 1847 and used as a resort until early in this century, are thought to be related to the recent volcanic activity in the Clear Lake area to the northeast.

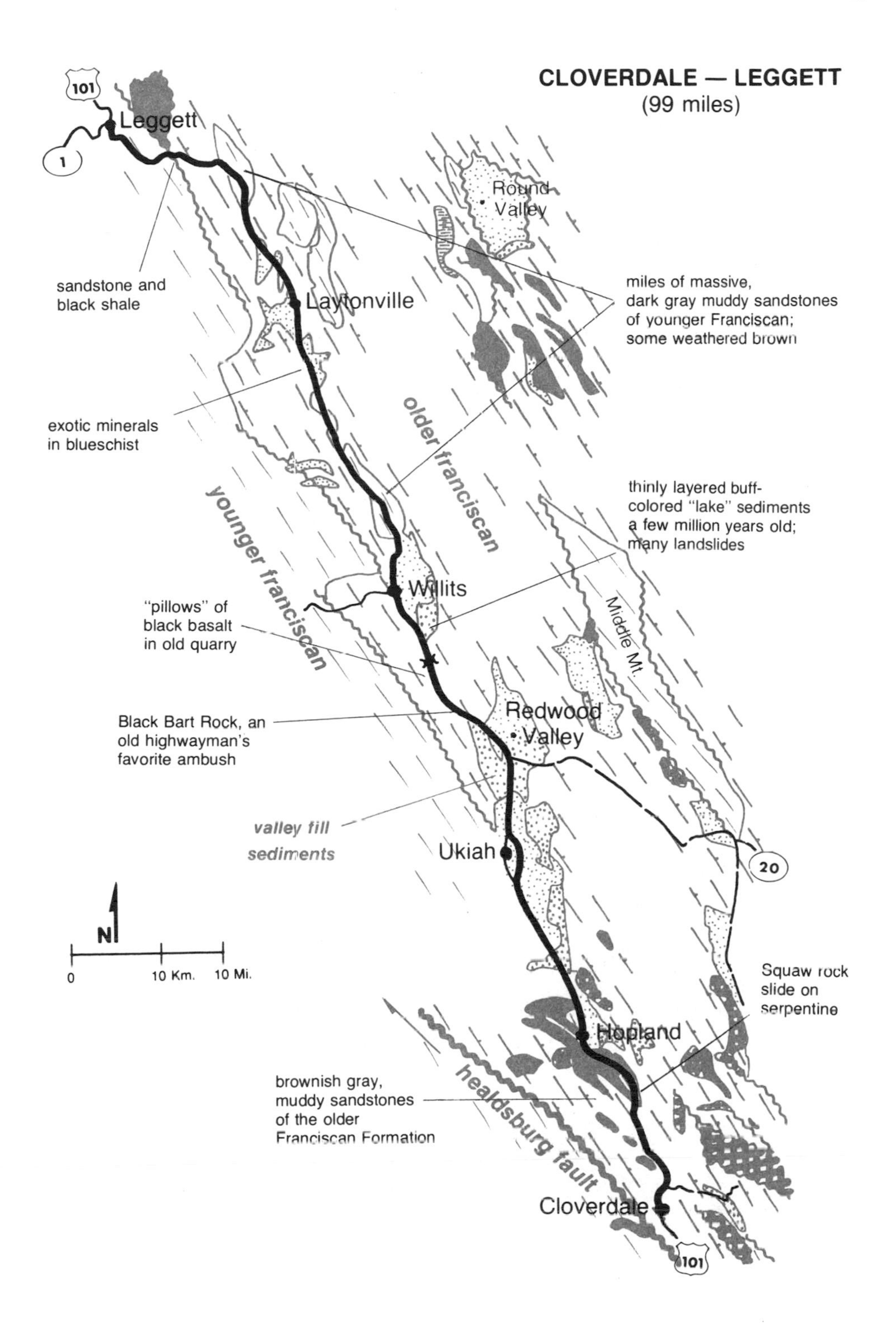
CLOVERDALE — LEGGETT
(99 miles)
101
1
Leggett
sandstone and
black shale
Round Valley
Laytonville
miles of massive,
dark gray muddy sandstones
of younger Franciscan;
some weathered brown
exotic minerals
in blueschist
older franciscan
younger franciscan
thinly layered buff-
colored "lake" sediments
a few million years old;
many landslides
Willits
"pillows" of
black basalt
in old quarry
Middle Mt.
Black Bart Rock, an
old highwayman's
favorite ambush
Redwood Valley
valley fill
sediments
Ukiah
20
N
0
10 Km.
10 Mi.
Squaw rock
slide on
serpentine
Hopland
brownish gray,
muddy sandstones
of the older
Franciscan Formation
healdsburg fault
Cloverdale
101

CLOVERDALE — LEGGETT

The long road between Cloverdale and Leggett winds through wooded hills along the rugged spine of the Coast Range. All of the rocks along the way are Franciscan — muddy and sandy sediments deposited on the deep ocean floor and then jammed against the continental margin. In places there are remnants of the sea floor itself.

Most of the old sea floor is between Cloverdale and Ukiah where there are outcrops of very dark rock, mostly slippery-looking dark green serpentinites. Soils developed on these rocks tend to be harshly infertile so the hills eroded in them are typically sparsely covered by a straggly growth of stunted shrubs that cloak the dark soil in patches and leave it exposed elsewhere. Rocks exposed between Ukiah and Leggett are almost entirely dark, muddy

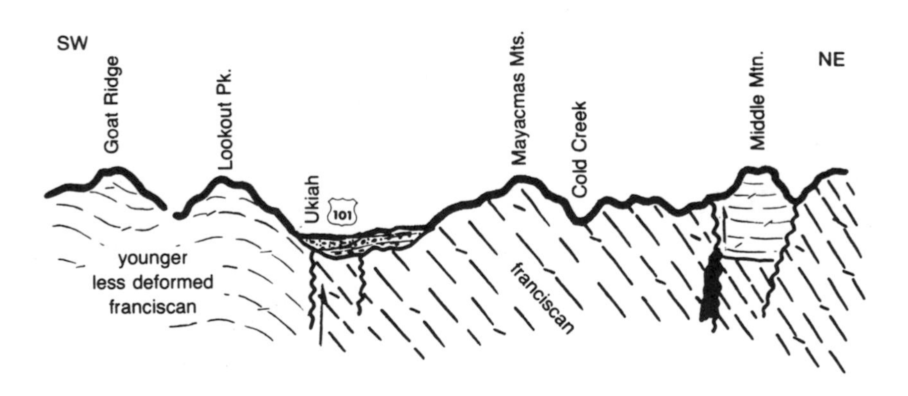

Section across U.S. 101 near Ukiah. The landscape in this area is eroded into Franciscan rocks.

sandstones that were folded and crushed as they were rudely jammed against the continent about 100 million years ago. Some of these rocks, especially those near Leggett, are distinctly less deformed than most Franciscan sediments — evidently they were among the last to be jammed into the trench.

The highway passes through a series of valleys between Cloverdale and Leggett. Ukiah is in the middle of the largest of these, Willits and Laytonville are in two smaller ones. These valleys almost certainly developed during the last few million years as a result of crustal movements along large faults parallel to the San Andreas. Their floors are deeply filled with stream sediments washed in from the surrounding mountains, porous deposits that form excellent agricultural soils with good subsurface drainage.

Pillows of basalt erupted under water. Quarry north of Ukiah.

101

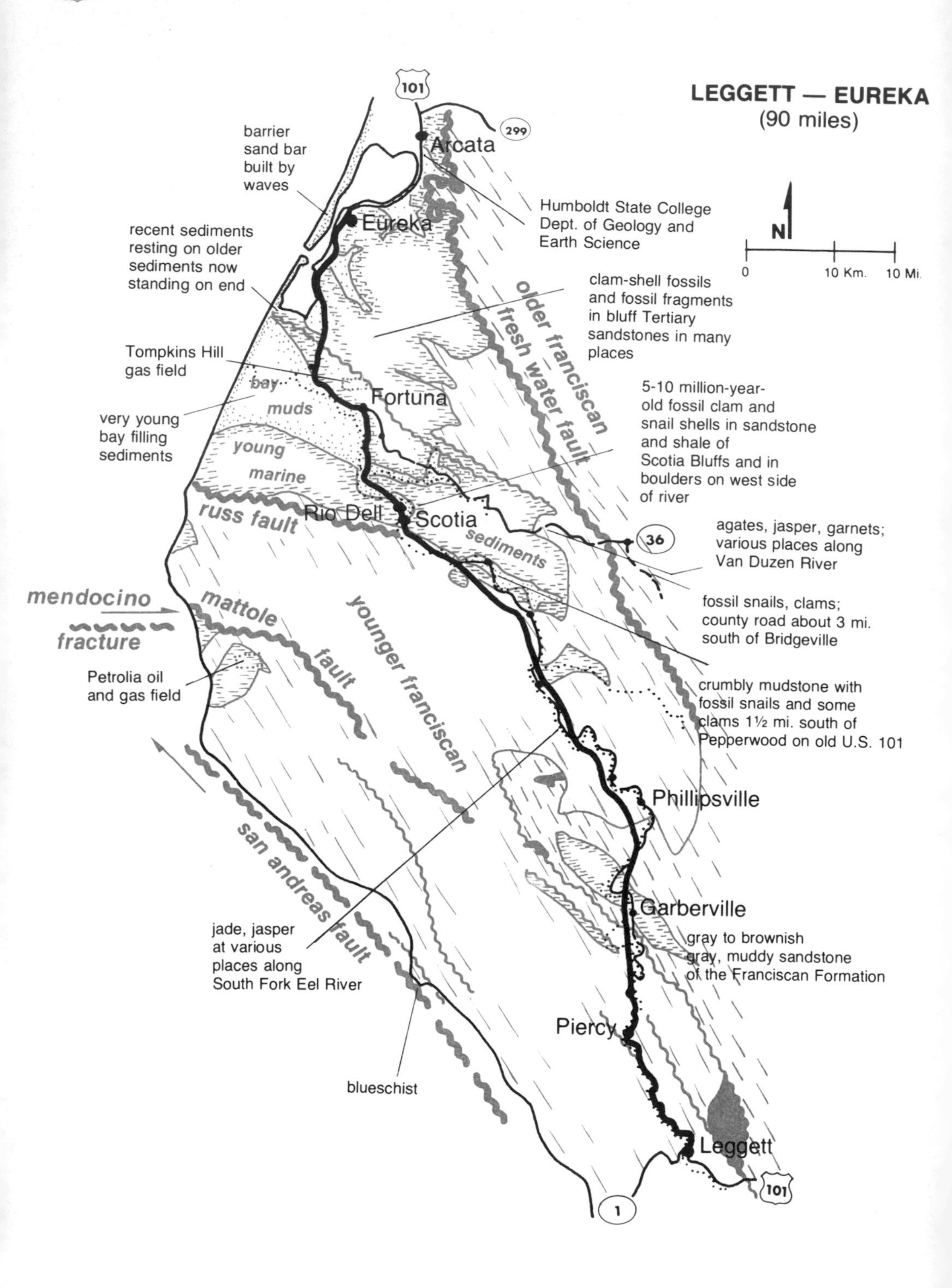
LEGGETT — EUREKA
(90 miles)
N
0
10 Km.
10 Mi.
101
299
Arcata
Eureka
Fortuna
Rio Dell
Scotia
36
Phillipsville
Garberville
Piercy
Leggett
101
1
barrier
sand bar
built by
waves
Humboldt State College
Dept. of Geology and
Earth Science
recent sediments
resting on older
sediments now
standing on end
clam-shell fossils
and fossil fragments
in bluff Tertiary
sandstones in many
places
Tompkins Hill
gas field
very young
bay filling
sediments
5-10 million-year-
old fossil clam and
snail shells in sandstone
and shale of
Scotia Bluffs and in
boulders on west side
of river
agates, jasper, garnets;
various places along
Van Duzen River
fossil snails, clams;
county road about 3 mi.
south of Bridgeville
crumbly mudstone with
fossil snails and some
clams 1½ mi. south of
Pepperwood on old U.S. 101
Petrolia oil
and gas field
gray to brownish
gray, muddy sandstone
of the Franciscan Formation
jade, jasper
at various
places along
South Fork Eel River
blueschist
bay
muds
young
marine
sediments
older franciscan
fresh water fault
russ fault
mendocino
fracture
mattole
fault
younger franciscan
san andreas fault

Franciscan sandstone and mudstone in roadcut 6 miles south of Leggett.

LEGGETT — EUREKA

Along most of the route between Leggett and Rio Dell the Redwood Highway follows deep and winding canyons. Numerous large outcrops make steep bluffs along the rivers. Occasional smaller ones higher on the hillsides mark old landslide scars where the soil has slumped off the slope.

Conspicuous outcrops between Leggett and Garberville, in the narrow canyon of the South Fork of the Eel River, expose layers of sedimentary rocks, mostly muddy gray sandstones deposited on the deep floor of the Pacific Ocean. Scattered fossil localities in this northern part of the Coast Range have yielded remains of animals a bit younger than those found in Franciscan rocks elsewhere and the outcrops expose rocks less brutally deformed. Many of the sedimentary layers visible from the road are only gently tilted, not crumpled and broken. Probably these were among the last sea floor sediments stuffed into the marginal trench during its final stages of activity — they escaped intense deformation because no more rocks were jammed in on top of them.

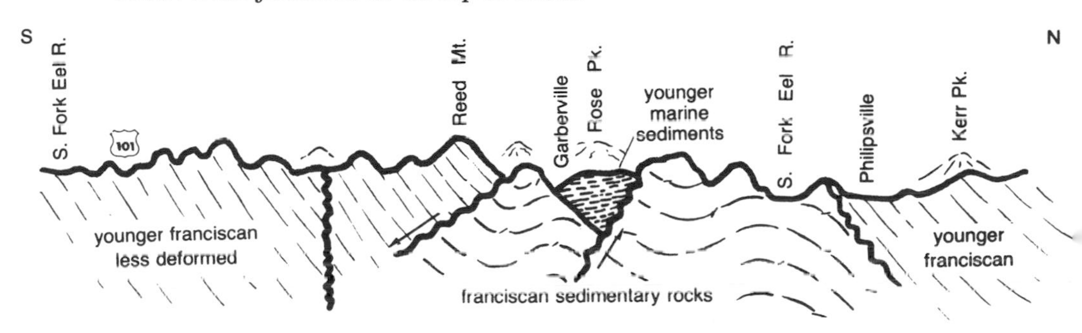

Section along U.S. 101 between Leggett and Eureka.

Near Garberville the South Fork of the Eel River flows for several miles through a broad valley it has eroded into very soft rocks. North of Garberville outcrops become sparse and inconspicuous, rarely noticed by travellers whose attention is riveted to the more spectacular redwood trees. Most of the few rocks that are visible are similar to those between Garberville and Leggett. Between the community of Holmes, off the new road on the Avenue of the Giants, and Rio Dell the highway follows a broad valley the Eel River has eroded along the line of the Russ fault. Rocks north of this fault, including some of those near the road, are mostly soft muds deposited in shallow water since the rise of the Coast Range. Their softness explains the width of the valley.

Most of the route between Rio Del and Eureka is across the flat or very gently rolling surface of the Eel River Basin. All of the rocks along this stretch of road are soft muds and sands deposited in relatively shallow water long after formation of the Coast Range. Fossil seashells are abundant in these muds and sands and are very easy to collect, but unfortunately good outcrops are hard to find.

Fossil clam shells in soft sandstone from the Scotia Bluffs.

The Eel River basin is a mud-filled trough bounded by faults that trend nearly east-west, parallel to the great Mendocino fracture on the floor of the Pacific Ocean just a few miles to the south. At the east end of the basin they bend around to the south becoming parallel to the San Andreas fault. It seems very likely that these faults are fractures in the continental crust caused by movement along the Mendocino fracture where it joins the San Andreas fault. Their present position a few miles north of the Mendocino fracture is probably the result of northward movement of the Coast Range along faults parallel to the San Andreas which ends at the Mendocino fracture just offshore from Cape Mendocino.

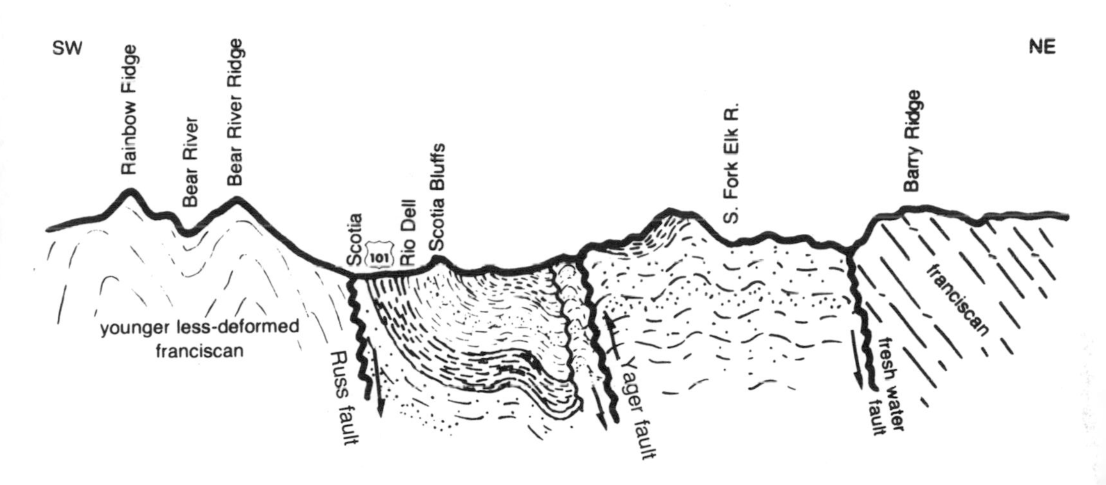

Section across the Eel River basin.

Any thick accumulation of muddy sediments deposited in seawater is a likely source of crude oil so the Eel River basin has been a target of wildcat drilling for many years. It looks especially tempting because other rather similar basins in the southern California Coast Ranges have produced large quantities of oil. But no significant oil fields have been found in the Eel River basin and hope fades dimmer with every new dry hole.

Eureka is at the narrows between Humboldt Bay to the south and Arcata Bay to the north. Both are drowned mouths of river valleys separated from the ocean by barrier bars, long sandspits built by the waves as they sweep sand southward along the beach. Until recently there was a similar bay at the mouth of the Eel River, just south of Humboldt Bay, but it is completely filled with mud carried in by the river. Nothing remains of this bay but broad, flat fields skirted by U.S. 101 between Scotia and Loleta about 15 miles south of Eureka.

101

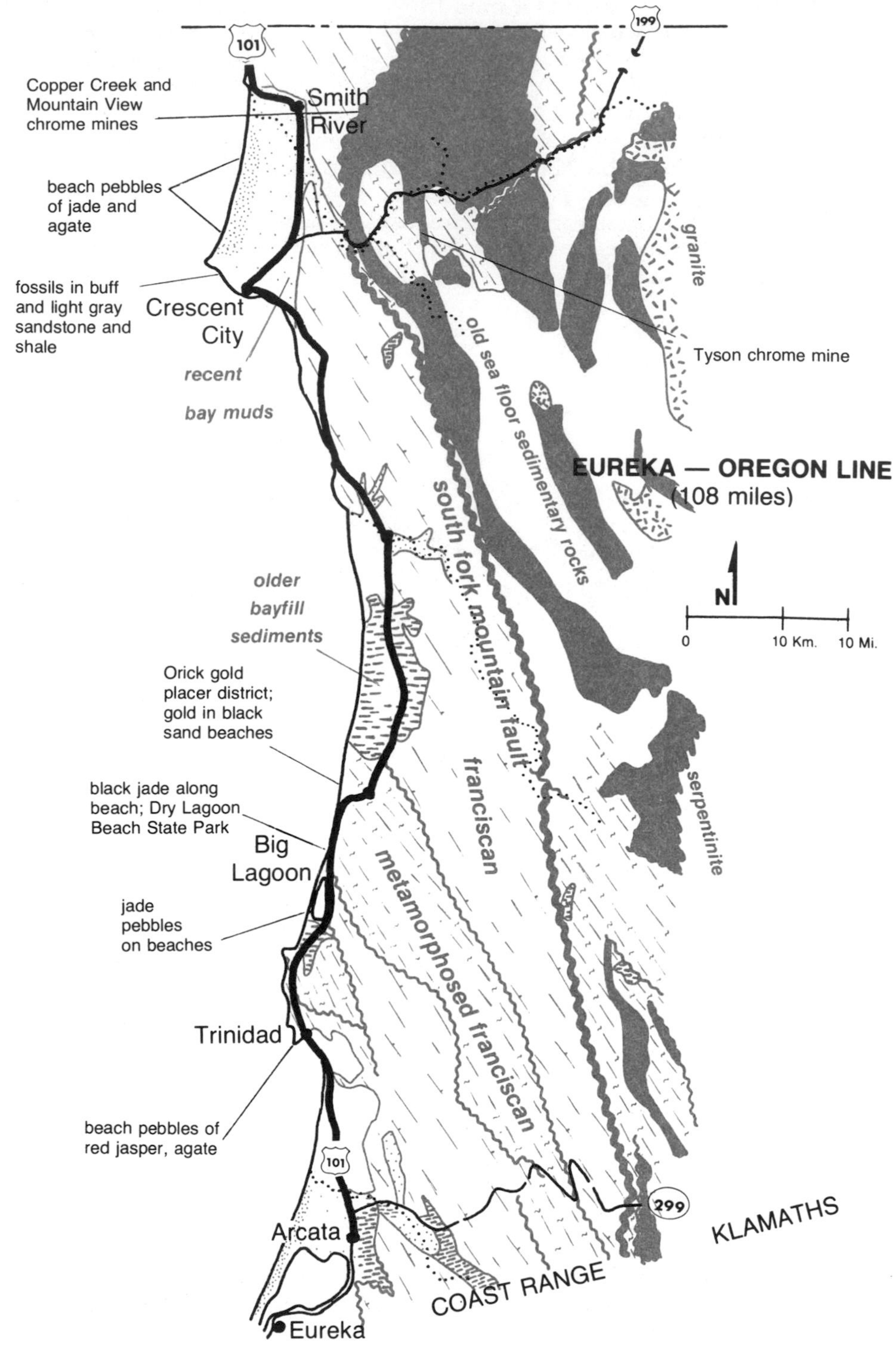
101
199
Copper Creek and Mountain View chrome mines
Smith River
beach pebbles of jade and agate
fossils in buff and light gray sandstone and shale
Crescent City
recent bay muds
granite
Tyson chrome mine
old sea floor sedimentary rocks
EUREKA — OREGON LINE
(108 miles)
N
0
10 Km.
10 Mi.
south fork mountain fault
older bayfill sediments
Orick gold placer district; gold in black sand beaches
black jade along beach; Dry Lagoon Beach State Park
Big Lagoon
franciscan
serpentinite
metamorphosed franciscan
jade pebbles on beaches
Trinidad
beach pebbles of red jasper, agate
101
299
Arcata
KLAMATHS
COAST RANGE
Eureka

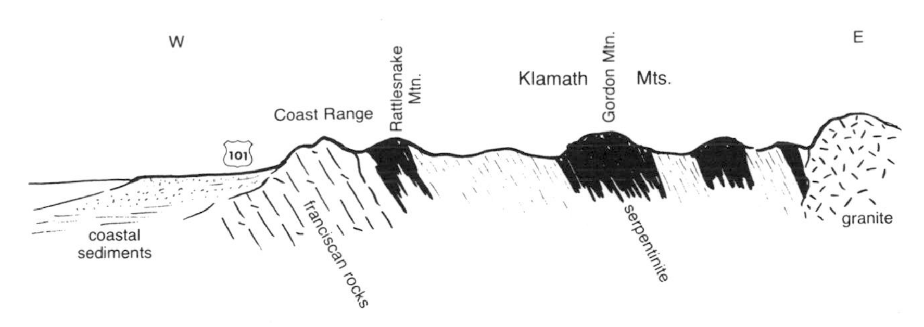

Section across U.S. 101 near Crescent City. The Coast Range is here reduced to a narrow fringe of Franciscan rock fringing the Klamaths.

EUREKA — OREGON LINE

The Redwood Highway stays within sight and smell of saltwater along most of the route between Eureka and the Oregon line, about 110 miles. This is the northernmost end of the California Coast Range where it narrows to an elongate panhandle of Franciscan rocks, as little as 10 miles wide, between the Klamath Mountains and the sea.

All of the rocks exposed along the way are Franciscan sediments, all muds and sands deposited far from shore on the deep floor of the Pacific Ocean and then stuffed into the marginal oceanic trench about 100 million years ago. Soils are deep in this rainy and well-forested area so there are few good exposures of bedrock near the road and those don't often distract attention from the giant redwoods. But there are excellent exposures all along the seacliffs and people who stop for a walk on the beach will clamber over wave-battered exposures of dark muddy sandstones.

This northern stretch of California's coast is distinctly different from that south of Cape Mendocino. Fog-shrouded hills clad in somber evergreen forests rise directly from the beach unbroken by the emergent marine terraces so conspicuous farther south along the Coast Road. Because there are no terraces to provide an easy and continuous route along the coast, the road winds through forested hills, onto the coast for a few miles, and then back into the hills again. Coastal lagoons, like those between Trinidad and Orick, are another feature of this northern coast that does not appear south of Cape Mendocino. They are the drowned mouths of stream valleys separated from the sea by wave-built sandspits. Evidently this part of the California coast is sinking instead of rising like the coast from Cape Mendocino southward. Perhaps this is because sea floor offshore to the north has been descending into the earth's interior pulling the coast down and jamming sediments into a marginal trench so stuffed that it does not show as an especially deep place on maps of the sea floor.

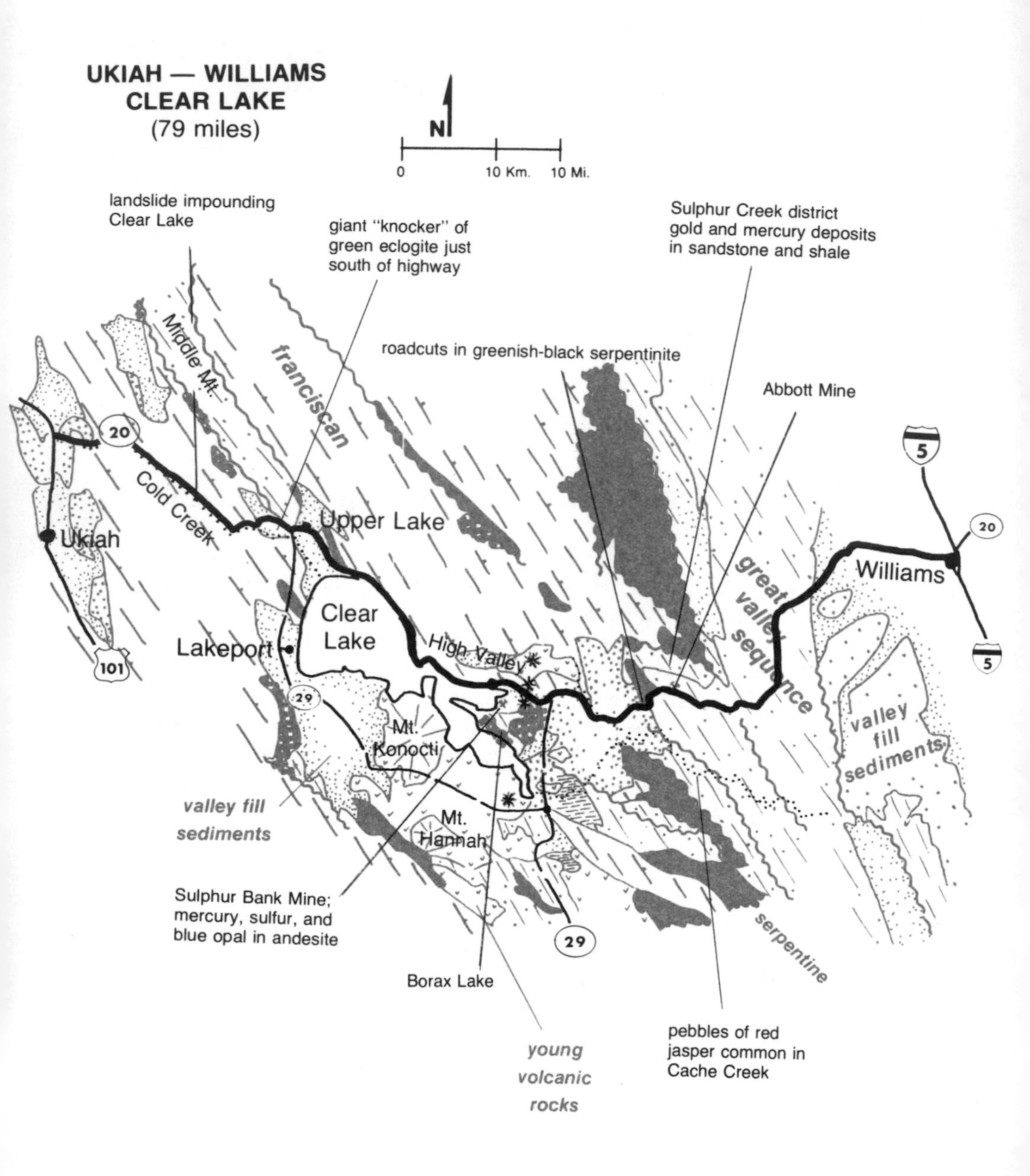

UKIAH — WILLIAMS
CLEAR LAKE
(79 miles)
N
0
10 Km.
10 Mi.
landslide impounding Clear Lake
giant "knocker" of green eclogite just south of highway
Sulphur Creek district gold and mercury deposits in sandstone and shale
roadcuts in greenish-black serpentinite
Abbott Mine
Middle Mt.
franciscan
20
Cold Creek
Ukiah
Upper Lake
Clear Lake
Lakeport
High Valley
Mt. Konocti
Mt. Hannah
101
29
great valley sequence
Williams
5
20
valley fill sediments
valley fill sediments
Sulphur Bank Mine; mercury, sulfur, and blue opal in andesite
Borax Lake
serpentine
young volcanic rocks
pebbles of red jasper common in Cache Creek

california 20

U.S. 101 — WILLIAMS

The junction with U.S. 101, about 8 miles north of Ukiah, is in the north end of the Redwood Valley, a broad depression within the Coast Range almost certainly created by movement along faults parallel to the San Andreas. Valley fill sediments deposited in the floor of the basin sometime within the last several million years are now carved by erosion into a gentle landscape of softly rolling hills.

Rugged hills surrounding the valley are underlain by much sterner rock, mostly dark Franciscan sandstones which include a few layers of muddy shale and other sediments. All the mountains between U.S. 101 and Clearlake Oaks are made of Franciscan rock.

Like most natural bodies of fresh water, Clear Lake owes its existence to a series of geologic accidents. It floods a broad valley that originally drained westward through Cold Creek into the Russian River until a large landslide plugged the drainage sometime within the last few thousand years. Water impounded behind the landslide filled the valley until it spilled through a new outlet into Cache Creek which drains eastward from the southeastern tip of the lake into the Sacramento River. Clear Lake still exists today because its outlet stream happens to flow over very hard rock, a lava flow, and is having difficulty eroding its channel deep enough to drain the lake. Had the spillway been established over the broken and easily erodible rock of the landslide dam, it would have drained the lake long ago.

Most of the landscape along the south and east sides of Clear Lake was produced by volcanic activity within the last few million years, some of it within the last few thousand years. Mt. Konocti dominates the scene on the south shore of the lake. It is a large volcano with several peaks which still retain most of their original form even though erosion has begun to leave its mark during the thousands of years that have passed since the last eruption.

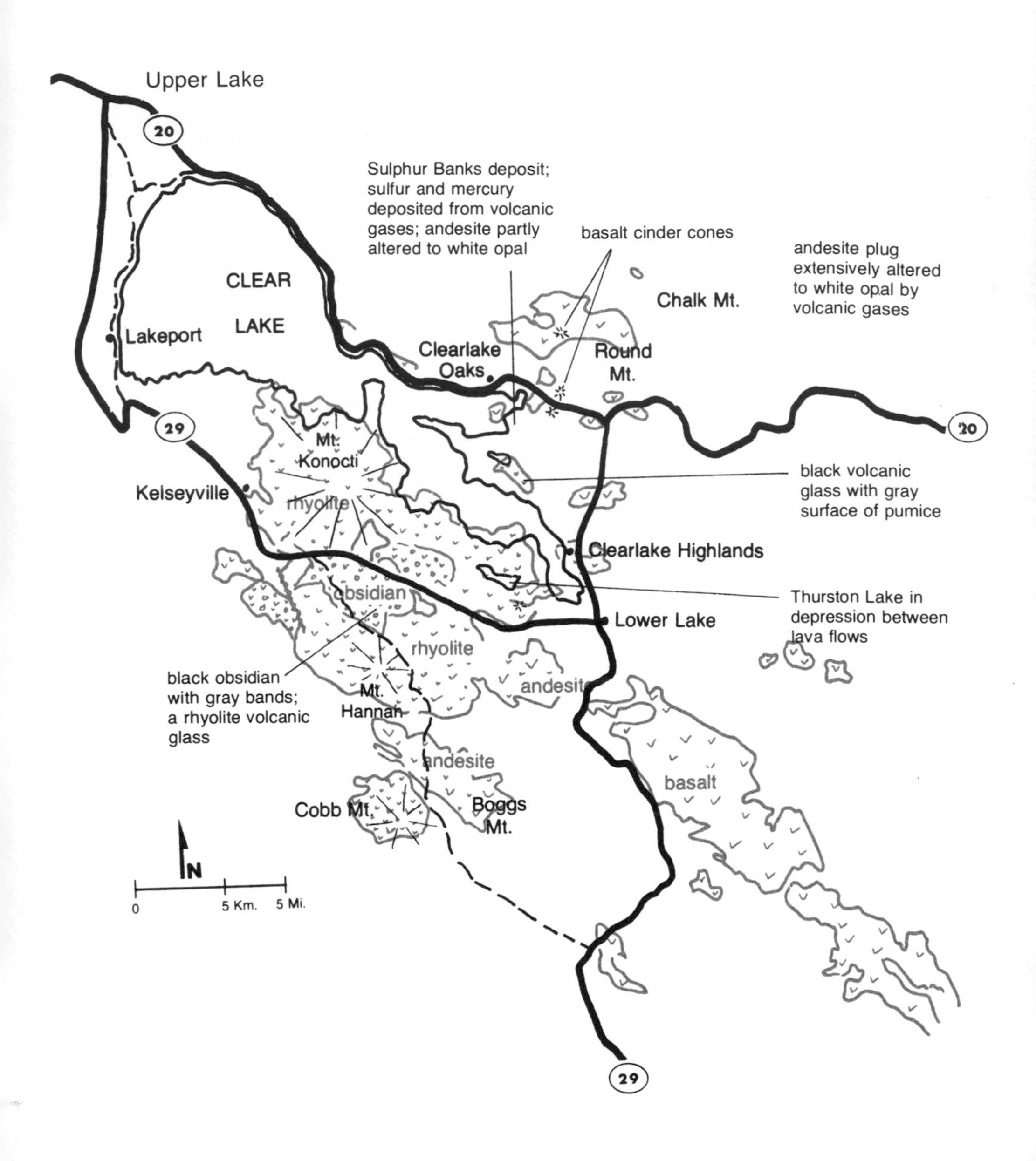

Sketch map showing young volcanic rocks in Clear Lake Area.

About two miles east of Clear Lake Oaks, Highway 20 passes between a matched pair of small cinder cone volcanoes, each about a mile in diameter and 400 feet high. Both are nearly untouched by erosion but weathering has reddened the basalt cinders by converting some of their mineral content to red iron oxides. It is difficult to guess the age of basalt cinder cones because they are surprisingly resistant to erosion; rainwater soaks into their loose surfaces so easily that very little runs off to erode gullies.

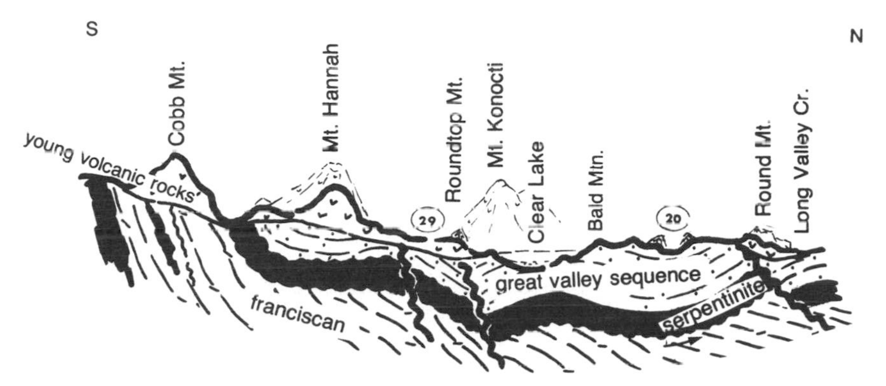

Section across the Clear Lake area showing young volcanic rocks perched on top of older hills eroded into Franciscan rock.

About a mile south of the paired cinder cones is the big Sulphur Bank lava flow which appears to be very young. Hot springs and gas vents are still active in its southern part which evidently covers the old volcanic mouth. The hot fluids have had their effect on the lava flow, converting much of it to white opal and leaving deposits of sulphur and mercury that have been mined intermittently for the last century. It is common to find ore deposits in rocks that have been altered by circulating hot water and mining geologists always watch for the pale pastel shades of such rock.

The eastern end of High Valley, about two miles north of the paired cinder cones beside Highway 20 and also accessible by side roads, is another area of recent volcanic activity. A nifty little cinder cone about 400 feet high squats in the midst of a jagged expanse of fresh lava still not covered by soil and supporting only a straggling few adventurous plants. Another neat little lava field is just south of Borax Lake, near the community of Clearlake Park, where there are flows of obsidian, a beautiful black volcanic glass. There is no volcanic peak here. The viscous lava squeezed slowly out of the vent and spread thickly, like so much runaway bread dough, over the surrounding area. A rounded hummock about 40 feet high near the center of the flow probably marks the location of the vent.

Wherever very recent volcanic rocks such as those around Clear Lake exist, there is always the possibility of new eruptions. And it is almost certain that extremely hot rocks exist within a few thousand feet of the surface, providing the potential for development of more natural steam generating plants such as the one that has been operating for many years at the Geysers, several miles southwest of Clear Lake. Hot springs actively depositing ore minerals, such as those at Sulphur Bank, are another encouraging sign. They show that deep circulation of hot water already exists and that it should be possible to find good locations to drill steam wells.

East of Clear Lake, Highway 20 winds for several miles through low hills underlain by loose sands and gravels eroded from the nearby hills and deposited in the floor of a broad valley during the last several million years. Approximately 12 road miles east of Clear Lake are a number of good road cuts in dark green serpentinite. Nearby hills underlain by this rock have a distinctly greenish look in places and are sparsely covered by plants. This is the belt of serpentinite that separates the Great Valley Sequence from the Franciscan — it was once the bedrock seafloor on which the muddy sands of the Great Valley Sequence were deposited.

A number of large roadcuts beside Highway 20 give good views of the muddy sandstones that make the Great Valley Sequence. They aren't very pretty — rather dirty-looking brown sandstones that really do look like they were once layers of mud and sand on the floor of the Pacific Ocean. Their layers tilt steeply downward to the east.

Great Valley sequence rocks in a roadcut beside highway 20.

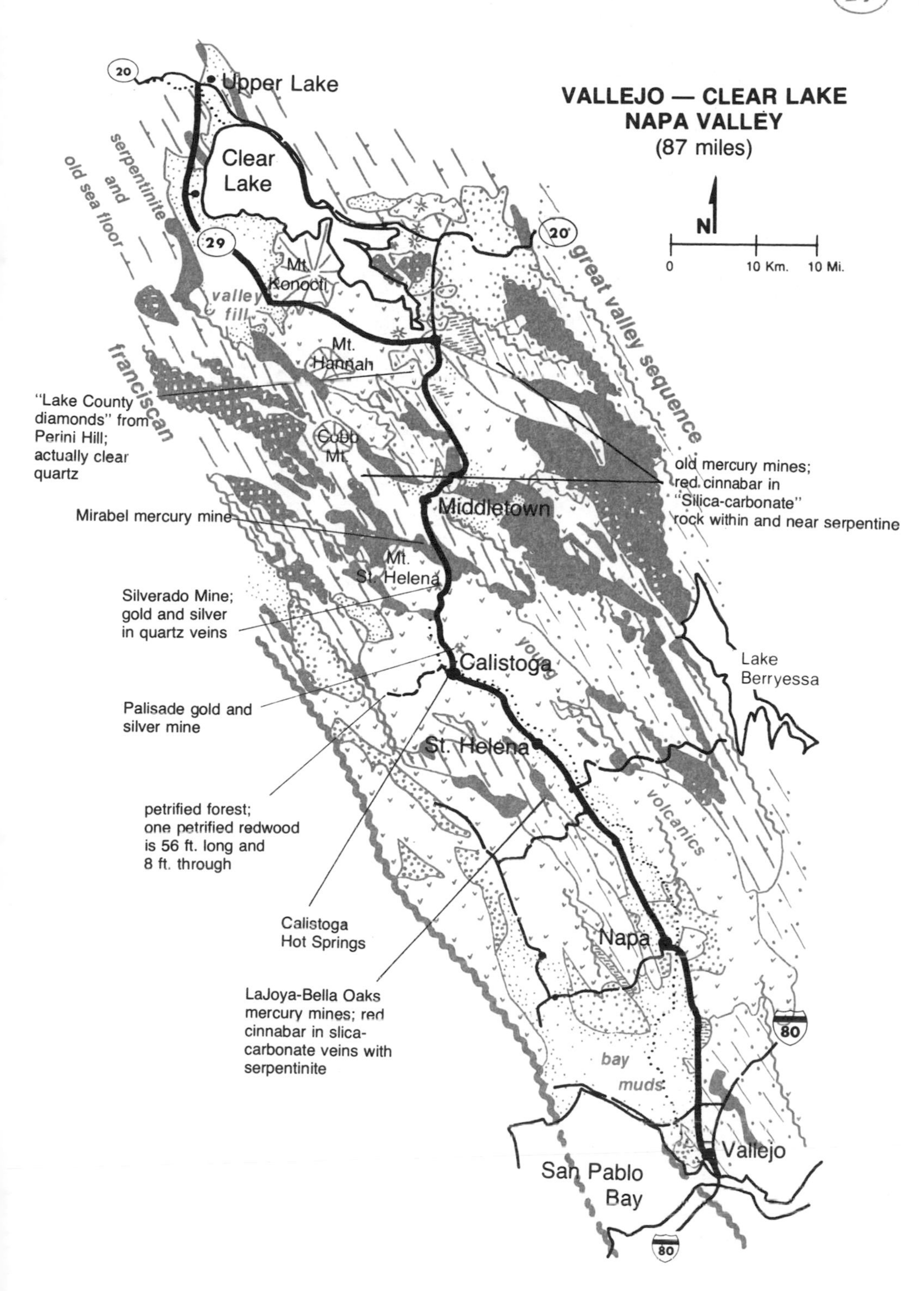
29
VALLEJO — CLEAR LAKE
NAPA VALLEY
(87 miles)
N
0
10 Km.
10 Mi.
20
Upper Lake
Clear Lake
serpentinite and old sea floor
29
Mt. Konocti
valley fill
20
great valley sequence
Mt. Hannah
franciscan
"Lake County diamonds" from Perini Hill; actually clear quartz
Cobb Mt.
old mercury mines; red cinnabar in "Silica-carbonate" rock within and near serpentine
Middletown
Mirabel mercury mine
Mt. St. Helena
Silverado Mine; gold and silver in quartz veins
Calistoga
young
Lake Berryessa
Palisade gold and silver mine
St. Helena
petrified forest; one petrified redwood is 56 ft. long and 8 ft. through
volcanics
Calistoga Hot Springs
Napa
LaJoya-Bella Oaks mercury mines; red cinnabar in slica-carbonate veins with serpentinite
80
bay muds
Vallejo
San Pablo Bay
80

california 29 — the napa valley

VALLEJO — CLEAR LAKE

Between Vallejo and Calistoga the route follows the floor of the Napa Valley, one of the larger of the numerous elongate valleys that trend northwest-southeast through this part of the Coast Range. They parallel the San Andreas fault and it seems likely that they were created by movements along the swarm of lesser faults belonging to the San Andreas system. These cut the Coast Range into a series of long slices that slid past each other, just as the opposite sides of the San Andreas fault are moving today, to create the long narrow ridges and valleys that form the basic framework of the modern landscape.

The Napa Valley is deeply floored with deposits of loose sand and gravel washed in from the surrounding mountains after the valley had formed. Such accumulations are very porous and form the basis for the open and well drained soil best suited for growing grapes.

Valley soils on which the grapes are grown are eroded from variable mixtures of rocks in the old Franciscan sea floor and from the younger volcanic rocks which cover them in most places. Full-bodied red wines grown on soils rich in dark minerals from the Franciscan sea floor have a sharpness of taste that must be mellowed by aging. White wines are not so affected. But white wines grown on soils formed from serpentine in some of the side valleys take on a sharp, "acid" taste. Soils derived mostly from the abundant younger volcanic rocks, or from granites elsewhere, do not impart such taste characteristics.

Almost all of the hills visible from the floor of the Napa Valley are underlain by volcanic rocks erupted during the last 10 million or so years in which the San Andreas fault system has been active. Deposits of pale gray, pink and yellow volcanic ash predominate, light-colored volcanic rocks that record numerous violent outbursts of explosive volcanoes. This kind of volcanic activity is often associated with movement along major faults.

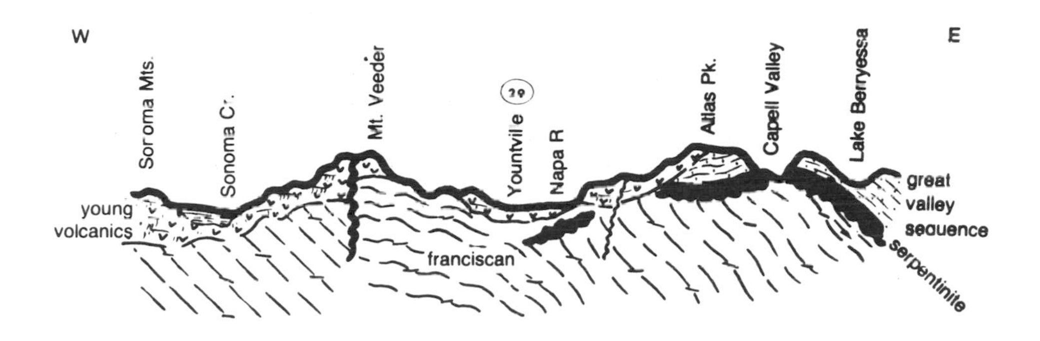

Section across California 29. Younger volcanic rocks cap the hills in much of this area.

Forests grew in these mountains while the volcanoes were active and many of them were buried in ash and then petrified as silica slowly soaked into the wood converting it to agate. Pebbles of petrified wood are common in many of the stream beds and petrified logs, even whole forests of petrified logs and stumps, have been found in a number of places in the hills. Many of the logs are very large — they conjure up a mental picture of lush forests growing on the slopes of volcanoes in a humidly subtropical climate. All this only a few million years ago.

Light colored volcanic rocks generally contain deposits of silver and those surrounding the Napa Valley are no exception. Old silver mines and prospects dot the hills, mementoes of the years when many of the mines produced impressive tonnages of good ore and Robert Louis Stevenson wrote of his sojourn at the Silverado Mine. Economic circumstances closed the mines, not lack of ore, and many of them will reopen when the price of silver gets high enough to make them profitable again.

Near Calistoga the road turns north to wind across the east flank of Mount St. Helena in a tortuous series of sharp turns and switchbacks. Roadcuts expose deposits of volcanic ash in a variety of pastel shades of gray, yellow and pink. Mount St. Helena, a complex of volcanic rocks, is not the remains of a simple volcano even though it looks like one. There have been no eruptions in this area for several million years and the old volcanoes are now so dismembered by movement along faults and carved by erosion that none remain as recognizeable elements of the landscape.

Just north of Mount St. Helena the road passes across a belt of serpentinite, darkly greenish rocks weathering to improverished reddish soils that offer grudging support to a scanty growth of scrubby trees and shrubs — digger pines are one of the few kinds of trees that flourish in such places. These and dark Franciscan sandstones, that weather to more hospitable soils, are exposed along the route between Mount St. Helena and Lower Lake.

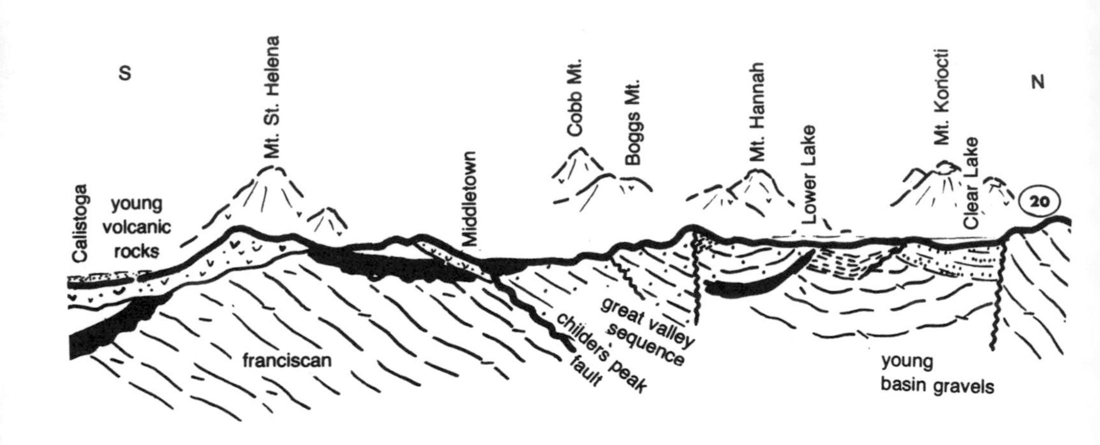

Section along the line of California 29 between Calistoga and Clear Lake. Most of the larger mountains are composed of volcanic rock.

Between Lower Lake and Lakeport, California 29 passes through the southern and western parts of the Clear Lake volcanic field, an area of very recent volcanic activity. Many of the hills and mountains along this stretch of the road actually are volcanoes so recently built that the processes of erosion have not yet had time to reshape them. Some of the lava flows are covered by soils and chapparal but others are still perfectly fresh; they must have erupted within the last few thousand years.

A "knocker" of blueschist in serpentinite.

These are mostly light-colored volcanics, like the other volcanic fields in the Coast Ranges, but they include large quantities of obsidian which is jet black. Its appearance gives no hint to the fact that obsidian has the same chemical composition as the light-colored ash with which it always associates. Obsidian is a glass stained black by its small content of dissolved iron. In the light-colored ash the iron is segregated into a few small crystals of black iron oxide sometimes visible as black specks. There is a lot of obsidian near the road along the stretch between Lower Lake and Kelseyville but most of it is well-covered by brush and difficult to find without a determined search. But obsidian is one of the most beautiful rocks and is always worth looking for, especially by people who need striking ornamental rocks for their gardens.

Between Lakeport and its junction with highway 20 at Upper Lake, California 29 passes through another stretch of hills eroded into Franciscan rock. These include some serpentinite exposed along the road near Rocky Point.

CROSS-REFERENCES

More about earthquakes and faulting in Chapter II, San Francisco Bay Area.

More about the Salinian Block in the introduction to Chapter II, The Coast Ranges.

More about landslides in Chapter II, San Francisco Bay Area.

More about the Clear Lake volcanic field in Chapter II, California 20: U.S. 101 — Williams.

Massive granites are the backbone of the Sierra Nevada.

III

sierra nevada and the klamaths

— a range divided

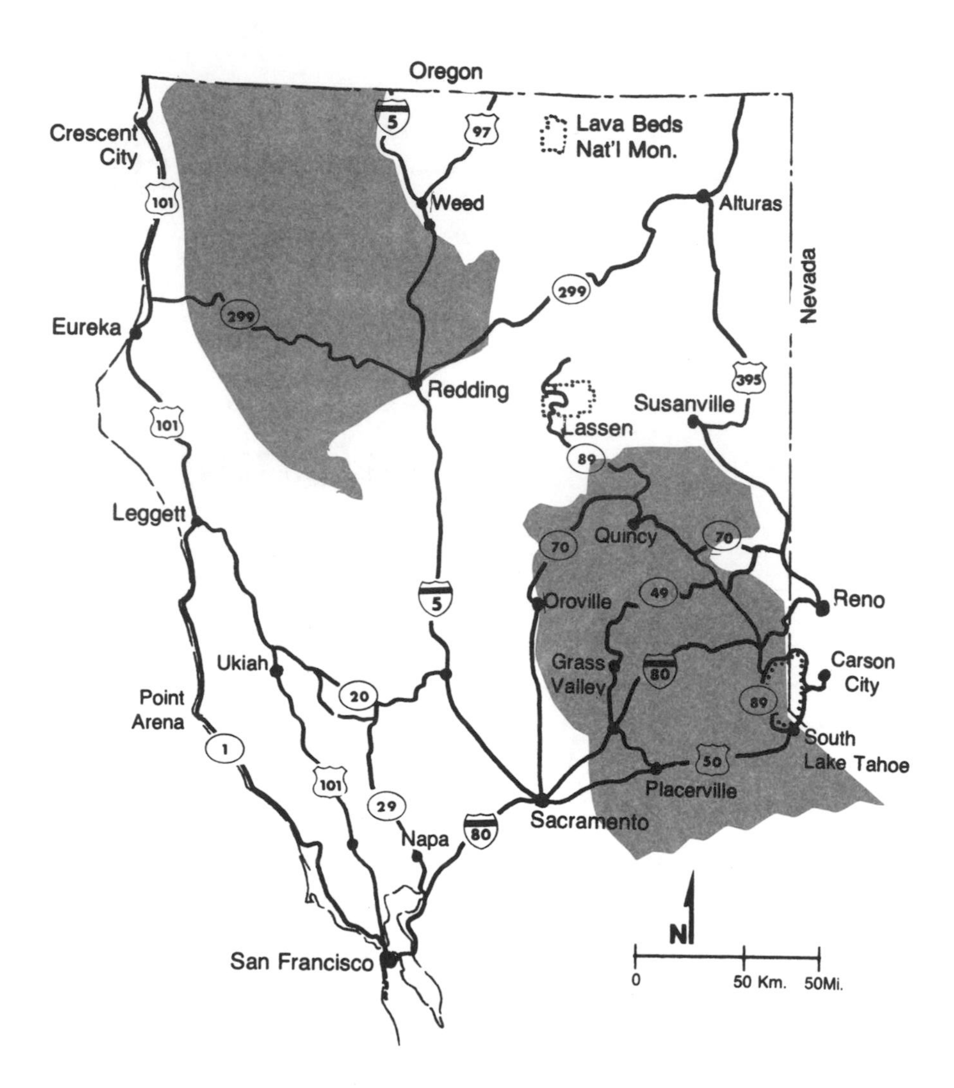
Oregon
Lava Beds
Nat'l Mon.
Crescent City
101
5
97
Weed
Alturas
Nevada
299
Eureka
299
395
Redding
Susanville
Lassen
101
89
Leggett
Quincy
70
70
5
Oroville
49
Reno
Ukiah
Grass Valley
80
Carson City
Point Arena
20
89
1
South Lake Tahoe
101
50
Placerville
29
Napa
80
Sacramento
N
San Francisco
0
50 Km.
50Mi.

— a range divided

California's eastern rampart, the Sierra Nevada, is shaped like a great wedge with a gentle western slope and steep eastern face. The glacially carved crags of the crest, near the eastern edge of the range, stand between the productive lowlands of the Great Valley and the parched deserts of Nevada. The Klamath Mountains tie a tangled northern knot between the Sierra Nevada and the Coast Ranges. Their higher slopes, clad in somber evergreen forests, rise to the jagged peaks of the Trinity Alps.

Geologists assumed for many years that the Sierra Nevada and Klamath Mountains were as distinct geologically as geographically but wondered at similarities in their rocks. Little was actually known because neither range had seen much geologic work and both are so heavily covered with soil and forests that their rocks are nearly invisible.

When new information about the geology of the northern Sierra Nevada and the Klamaths began to emerge during the 1960's, it quickly became quite obvious that the two are in fact separate parts of a single dismembered mountain range. Rocks along the southern edge of the Klamaths match those along the northern end of the Sierra Nevada almost as well as the type matches on pieces of a torn newspaper. Each major belt of rock, except one, and each major fault in the southern Klamaths has its counterpart in the northern Sierra Nevada. Evidently a single mountain range somehow broke into two pieces which moved apart about 60 miles.

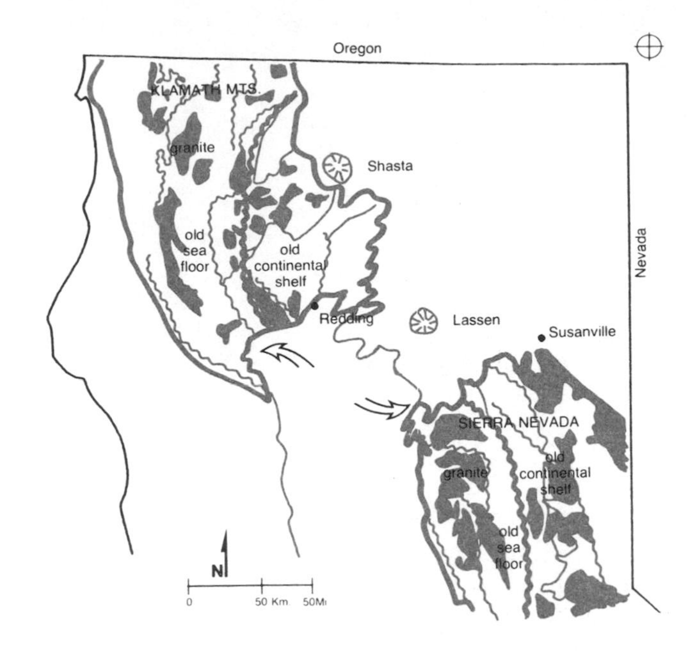

Curvature of rock belts near the matching edge shows how the range bent before it finally broke.

The best available evidence suggests that the Klamaths and Sierra Nevada separated about 140 million years ago when the rocks in the two ranges were still young and the processes that made them were still active. Sedimentary rocks containing fossils of animals that lived along shorelines during the middle of Cretaceous time, about 150 million years ago, are exposed in a few places along the western slopes of the Klamaths and the Sierra Nevada. Evidently they are remnants of a continuous shoreline that was broken when the ranges separated. Slightly younger sedimentary rocks lap onto the torn southern edge of the Klamaths where they could not have been deposited until after the ranges had broken apart.

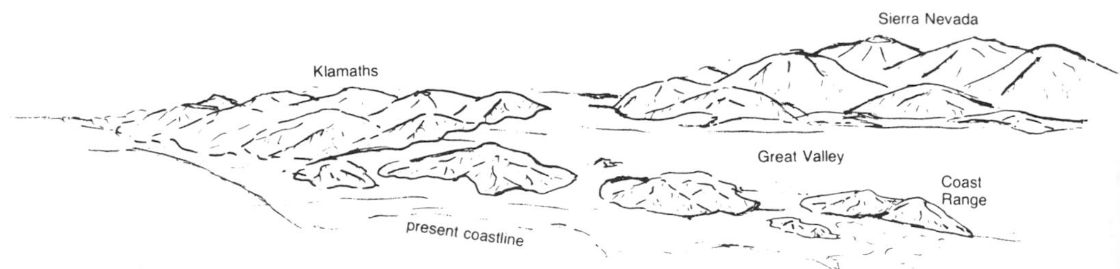

80 million years ago – volcanoes in the Sierra, sea water in the Great Valley, and islands along the coast.

The basic framework of both the Klamaths and the Sierra Nevada is a series of long belts of metamorphic rocks — old sedimentary rocks recrystallized by prolonged heating. After these had formed, they were intruded by enormous masses of granite that rose into them as molten magma. The record is clearest in the Klamaths because the old sedimentary rocks there are neither so thoroughly recrystallized nor so heavily intruded by granites as those in the Sierra Nevada. Apparently intrusion of granite magmas ceased in the Klamaths after the two ranges separated but continued for millions of years longer in the Sierra Nevada.

In both the Sierra Nevada and the Klamaths the belts of recrystallized sedimentary rocks are oldest in the east and become progressively younger westward. Large areas of the Klamaths are underlain by sheets of black igneous rocks, the kind that exist beneath the earth's crust. Similar rocks in the Sierra Nevada are more thoroughly changed by squeezing and reaction with water into belts of serpentinite resembling those in the Coast Ranges.

Actually there are also many important similarities between the rocks in the Sierra Nevada and Klamaths and those in the Coast Ranges. So many that we must conclude that all were formed in essentially the same way. Evidently the collision between continent and ocean floor began about 200 million years ago as near-shore sedimentary deposits were jammed onto the edge of the continent to create the Sierra Nevada and Klamath Mountains. As slice after slice of sea floor accumulated in the growing welt of the continental margin, the action moved steadily westward taking progressively younger sediments brought in from farther out on the sea floor.

Most of the muddy sediments swept against the continent from the moving sea floor are scraped off at the edge to make coastal mountain ranges. The heavier basalt under them rides on beneath the continent where it begins to melt to form magmas that float upward through the crust because they are lighter than solid rocks. Some of the crustal rocks also melt and mix with the original magma rising from beneath to form enormous masses of molten granite that go all the way to the surface where they fuel volcanoes. No doubt the Sierra Nevada and Klamath Mountains were once crowned by a chain of volcanoes similar to the modern Cascades. And we can imagine that the Cascades, which operate on magmas derived from rocks more recently swept under the crust, will leave granite batholiths to mark their roots when they finally erode away.

The granite magmas that intruded the Klamaths and Sierra Nevada formed during accumulation of the Coast Range. Similar granites never intruded the Coast Range because no younger mountains were built seaward; the mountain building process stopped. The presence of granite in the Klamaths and Sierra Nevada, and the cooking effects the granite magmas had on some of the sedimentary rocks already there, is the most important difference between these mountains and the Coast Range.

Large bodies of granite, called batholiths by geologists, seem at first to be extraordinarily monotonous expanses of granular pink and gray rock. But very careful mapping shows that they are composites of numerous intrusions, each several miles across, packed together like a bagful of marshmallows. Age determinations done by analysis of radioactive minerals show that these great bubbles of magma rise into the crust one by one during periods of intrusion lasting millions of years.

Wherever crushed and recrystallized sedimentary rocks are intruded by large masses of granite, as in the Sierra Nevada and Klamath Mountains, deposits of gold are almost certain to exist. Most often the gold is disseminated through veins of snowy white quartz that fill fractures opened in the sedimentary rocks as the masses of granite magma pushed and jostled their way upward. Circulating hot waters driven through the rocks by heat from the granite magma dissolve quartz and gold from large volumes of rock deep within the crust and then deposit them in concentrated veins near the surface where the water cools. At high temperatures gold and quartz seem to dissolve and deposit from water solutions under about the same conditions so we usually find them together.

Gold-bearing quartz veins form the "mother lode," eagerly sought by the early prospectors who imagined that it would surely make them incredibly rich. More of them were destined for an early grave than for wealth; quartz veins bankrupt hundreds of people for every one they enrich and most of those who work them barely manage to scratch out a squalid survival. Very few quartz veins are bonanzas; only a tiny minority contain any gold at all, usually no more than a trace. Even then it is almost always in widely scattered pockets or ore separated by enormous tonnages of barren white quartz. The gold is often so finely disseminated that it must be eroded and then transported in streams, nature's sluice boxes, to become concentrated enough to be worth mining. That is a later episode in the long history of the Sierra Nevada.

Platinum, as well as gold, has been mined in the streams of the Klamath Mountains. This rare mineral comes from the sheets of black igneous rock, old sea floor, that rode quietly over the top of the action while younger rocks were stuffed underneath them. Wherever large areas of such rocks are found the streams that have eroded them are almost certain to carry small amounts of platinum in their sands. Similar black igneous rocks, or serpentinites directly derived from them, in the Coast Ranges and the Sierra Nevada should also contain platinum but very little has yet been reported in those areas.

About 80 million years ago, while slices of sea floor were still jamming onto the Coast Range, activity ended in the Sierra Nevada and Klamaths and a time of relative geologic peace began. The last crushed slice of sea floor had jammed into the Coast Range and the last mass of molten granitic magma risen into the Sierra Nevada. Volcanoes that had been blowing steam and ash for millions of years snuffed out as the masses of molten magma hidden deep beneath them in the crust froze to become enormous bodies of granitic rock. The Sierra Nevada had long since separated from the Klamath Mountains which stood now, nearly an island, rising above sea water flooding the Great Valley and the area of northeastern California that is now the Modoc Plateau. Thus began a long period in which the slow processes of soil formation, erosion, transportation and deposition

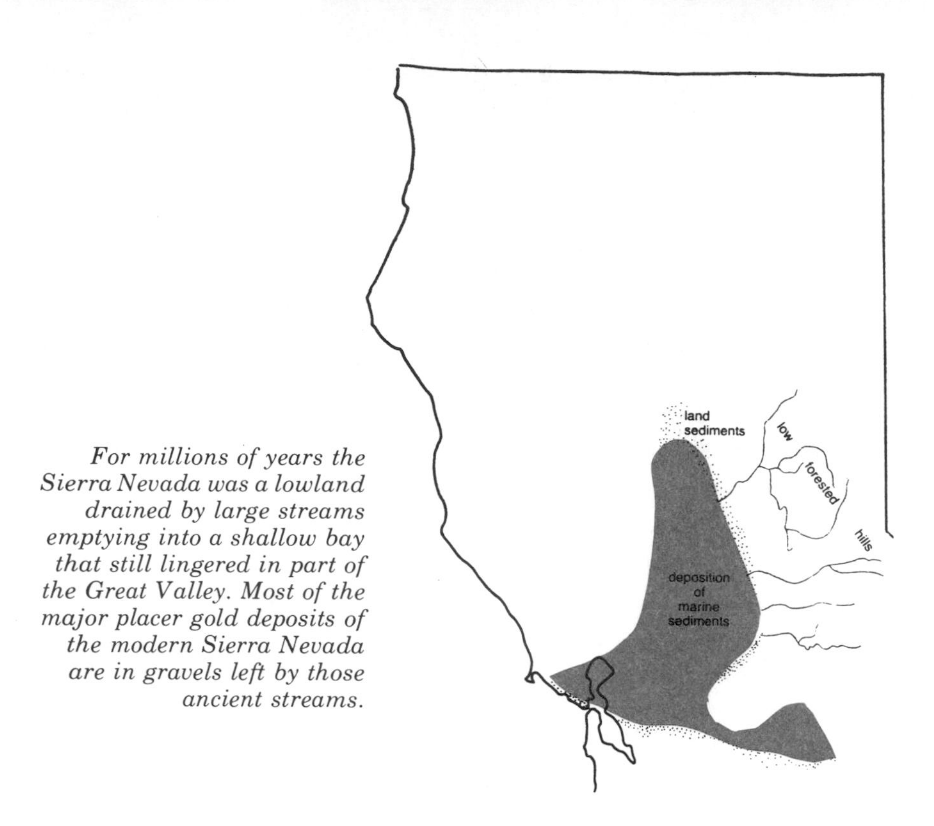

For millions of years the Sierra Nevada was a lowland drained by large streams emptying into a shallow bay that still lingered in part of the Great Valley. Most of the major placer gold deposits of the modern Sierra Nevada are in gravels left by those ancient streams.

of sediments played the major roles in developing California's modern landscape.

High and exposed to the erosive processes, vulnerable because they consist mostly of loosely-knit ash and rock fragments, the volcanoes must have melted away quite rapidly, shedding the muddy detritus of their decay into the inland seas. Then the rocks beneath them were stripped away until the enormous masses of granite and the complexly crushed and recrystallized sedimentary rocks that enclose them were exposed. All of this must have taken millions of years but it happened early in the long episode of quiet. By the time the Cretaceous Period ended and the Tertiary Period began, about 60 million years ago, most of northern California had probably been reduced to a rolling upland rising gently from inland seaways already largely filled with sediment.

During the next 40 or 50 million years, most of the Tertiary Period, northern California remained generally quiet. Erosion proceeded slowly in the land areas as their hills gradually melted away under a thickening blanket of soil and their rivers carried more sediment into seaways already nearly filled. Occasional changes of sea level, or perhaps minor vertical movements of the continent, caused the shoreline to shift back and forth sometimes flooding parts of the Sierra Nevada and other times converting the Sacramento Valley, now virtually filled with sediment, into dry land. It seems likely that the Klamath Mountains were a gently rolling upland, like the Sierra Nevada, during most of this long period.

The climate must have been warm and rainy during much of this long time, at least during the latter part of it. Red soils, laterites, that developed then remain today deeply blanketing large areas of the western foothills of the Sierra Nevada and parts of the Klamath Mountains. Such soils form wherever the climate is warm and abundant rainfall percolating downward through the ground dissolves and carries away everything but the least soluble substances, mostly aluminum oxide and red iron oxide. Fertilizer nutrients, being relatively soluble, are sadly lacking in lateritic soils so crops and garden plants need lavish applications of fertilizer to thrive. But many trees grow beautifully in these rain-impoverished soils so the forest industries prosper in such areas even though other agriculture may be marginal.

Serpentinites usually contain small quantities of nickel which frequently accumulates in lateritic soils developed on them. There are nickel mines in Oregon that operate in laterite ores and it seems likely that such deposits may also exist in California. Nickel laterites are mined simply by stripping the soft soil with ordinary excavating equipment.

Those millions of years of warm rain attacked the gold-bearing veins of the "mother lode" decomposing the quartz and releasing the gold, unchanged, into the soil and ultimately into the streams. Gold is very heavy so it works its way downward through stream gravels to bedrock where it lags behind and accumulates as lighter minerals are swept onward by floods. Weathering of bedrock to form soil releases gold from quartz, just as mining and milling does, and stream transportation concentrates it as though it were washed in a sluice box. It remains for the miner to recover the concentrated gold from the lowest parts of the stream gravels where it lodges behind irregularities of the bedrock as though they were riffles in the bottom of a sluice box. This is usually easier and more profitable than mining the original quartz veins, crushing the rock to mill the gold free, and washing it clean in sluice boxes. California's big placer gold deposits began their long history of accumulation during the quiet years of the Tertiary Period while the rocks in the Sierra Nevada and Klamath Mountains were weathering and eroding under a warmly humid climate.

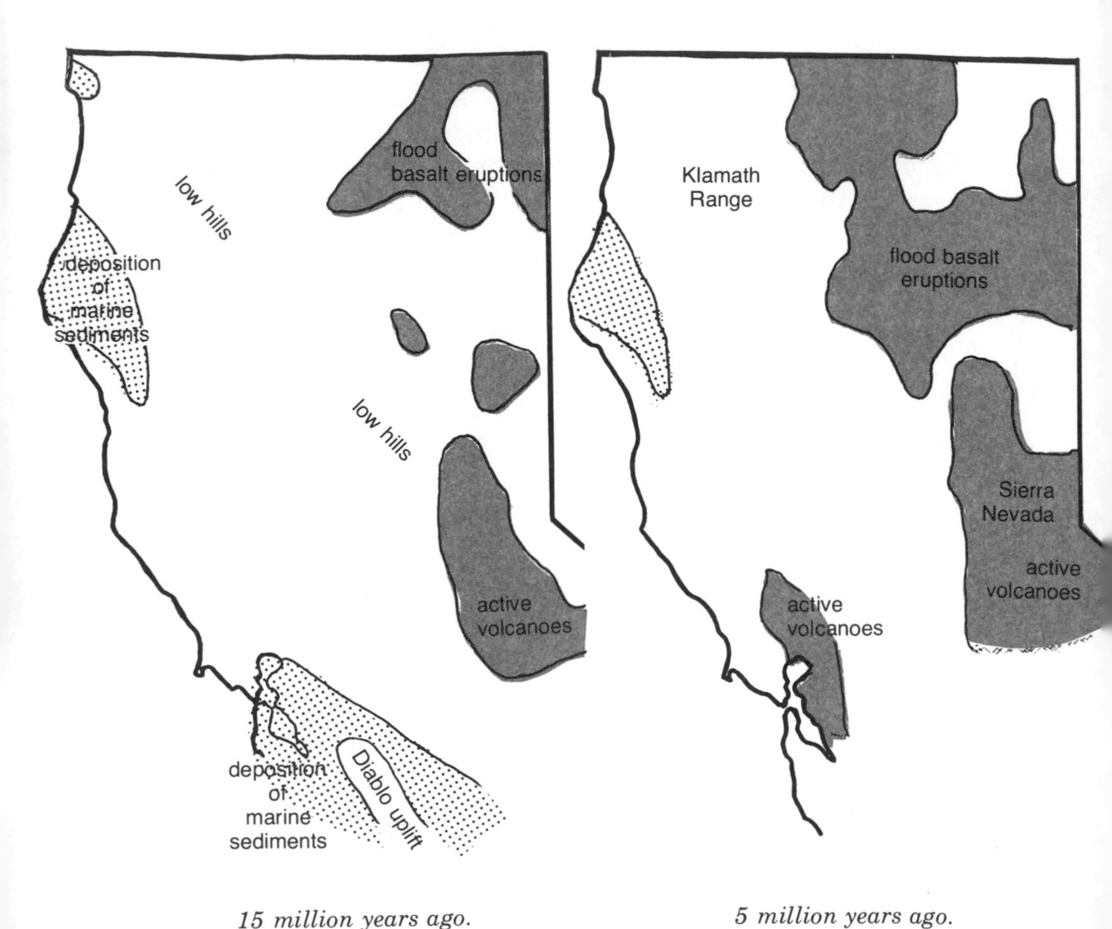

15 million years ago.

5 million years ago.

About 15 or 20 million years ago, the long period of geologic quiet in California ended. Action in the northeast began with a series of enormous floods of black basalt lava that covered the sedimentary rocks that had accumulated there while the area was a seaway and built a high tableland — the beginning of the Modoc Plateau. Simultaneous eruptions in the Sierra Nevada sent rivers of basalt lava down several major stream valleys sealing the gold-bearing gravels beneath thick lava flows and diverting the streams to other courses through the soft lateritic soils of the countryside. Now several of those old valleys stand high as meandering ridges of stream gravel protected from erosion by their caps of hard basalt. Oroville's Table Mountain is the best known of these.

Then a large region of the earth's crust, from the present Sierra Nevada eastward, not including the Klamath Mountains, began to stretch and break into large blocks. The largest and westernmost of these blocks is the Sierra Nevada. It rose along faults that define the east face of the present range, tilting the old land surface gently westward. Other blocks rose and fell east of the Sierra Nevada to make the desert ranges and basins of eastern California and Nevada.

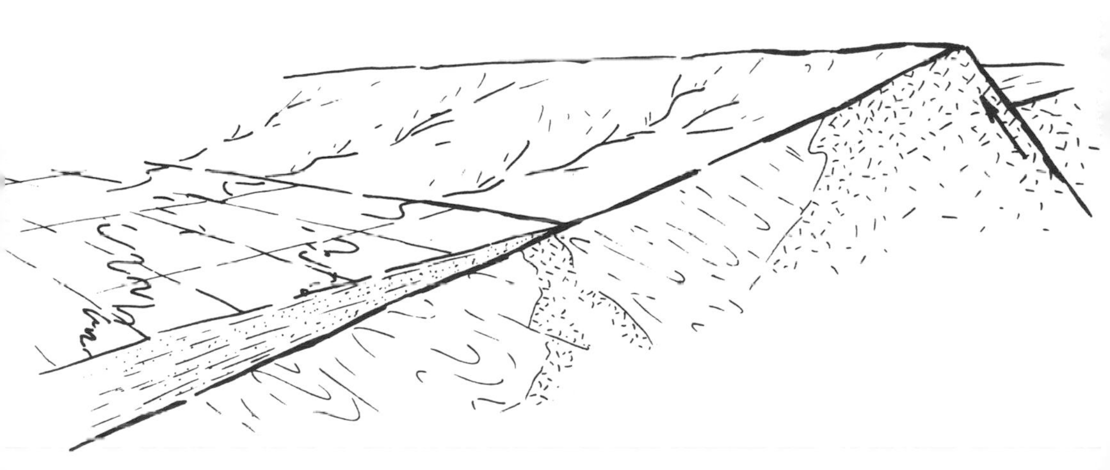

More volcanic eruptions producing basalt lava flows and clouds of volcanic ash accompanied the new movements of the earth's crust. These eruptions mantled large areas of the Sierra Nevada under blankets of white volcanic ash, poured more lava flows down stream beds, and built new volcanoes partially filling some of the block valleys. Similar activity occurred simultaneously in the Modoc Plateau. Most of the streams that had flowed over the Sierra Nevada during the millions of quiet years while the region was nearly flat and warmly humid, were now buried under volcanic deposits or diverted from their old courses as the land tilted.

New streams formed, eroding their new valleys across the abandoned channels of the old ones now mostly buried beneath blankets of volcanic debris. These are the streams that flow today, the ones in which prospectors first found placer gold. The oldtimers immediately began tracking the placer gold to its source by working their way upstream panning samples as they went. Most often they found the source in places where modern streams cut across the dry and abandoned channels of the old streams that had flowed during the millions of geologically quiet years. Their gravels were the dry placer deposits attacked by the hydraulic miners. Finding placer deposits of gold in the Sierra Nevada is still largely a matter of carefully tracing the old stream courses now partly buried under volcanic deposits and partly eroded away by the modern streams. This requires methodical geologic work. The electronic metal-finding gadgets widely sold to hopeful prospectors are absolutely useless for finding gold.

Uplift of the Sierra Nevada and development of the modern landscape are both far from complete. The fault that defines the eastern face of the range is still actively moving and occasional earthquakes jolt the crest a few feet higher. Some of the volcanic rocks along the eastern base of the Sierra Nevada are very recent and new eruptions would come as no surprise. The modern streams have only begun the long job of carving a new landscape into the old surface tilted upward by faulting just a few million years ago.

The Klamath Mountains seem to have been spared some of the geologic adventures experienced by the Sierra Nevada. We do not see in them the remains of an old lowland surface that was blanketed with volcanic debris and then uplifted to become renewed mountains. Instead we see the kind of rumpled landscape of ridges and ravines that forms during millions of years of erosion under a dense cover of vegetation. So the Klamath Mountains do have a different look and feel than the Sierra Nevada — mostly the result of differences in their later history instead of differences in their underlying bedrock skeleton.

Granite peaks south of U.S. 50 near Twin Bridges. Large slabs of weathered rock are flaking off the surfaces.

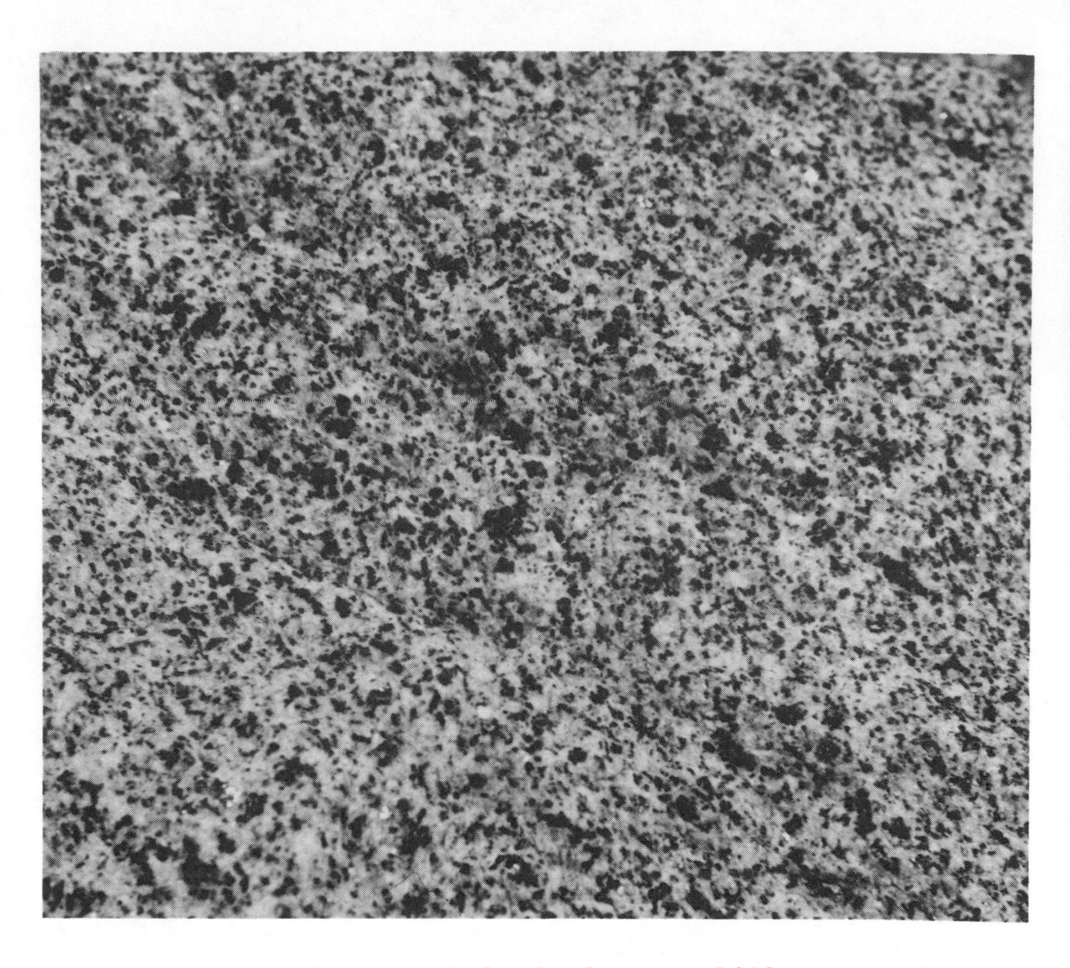

Granite as it looks close up. Light-colored quartz and feldspar peppered with black biotite and hornblende. About natural size.

Canyon carved into the old sierran surface by the American River.

interstate 80
sacramento — reno
— crossing the high sierra

Between Sacramento and Reno, a distance of 131 miles, the interstate highway passes over the high wedge of the Sierra Nevada which separates the lush lowlands of the Sacramento Valley from the bleak desert of Nevada. Much of the route crosses remnants of the gently rolling lowland that existed here before the Sierra Nevada was hoisted as though it were a flat cellar door hinged at its western edge. Red soils and dry stream gravels exposed along miles of the route across the western slope are relics of the old landscape now being remodelled by the modern streams. Ice-age glaciers carved the crest of the range exposing its granite core in bold outcrops of massive gray rock still nearly nude because the last ice-age ended so recently that there hasn't been time for a thick soil cover to develop. The steep eastern slope is the raw bedrock face of the Sierra Nevada fault still only slightly carved by streams and glaciers.

SACRAMENTO — EMIGRANT GAP
(74 miles)

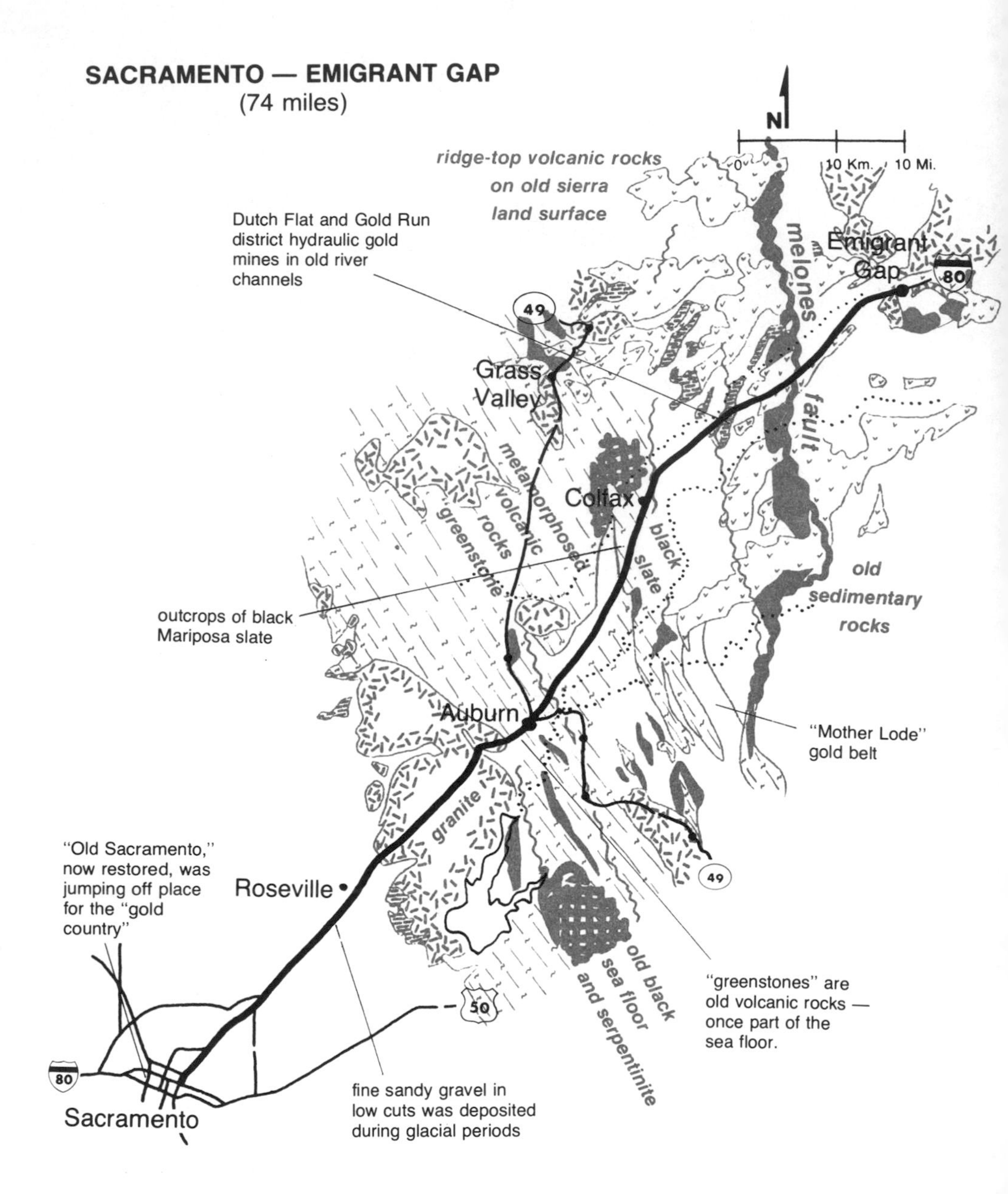

SACRAMENTO — EMIGRANT GAP

Between Sacramento and Roseville the highway crosses young alluvial fan deposits of mud, sand and gravel washed out of the canyons of the Sierra Nevada onto the floor of the Great Valley during the last several millions years. Roseville is on the thin edge of these deposits where they lap onto the eroded outcrop of a small body of intrusive granite. Bedrock outcrops between Roseville and Auburn are boulders of granite rounded where they stand by weathering.

The body of granite crossed by Interstate 80 between Roseville and Auburn is approximately circular in outline but its western edge is covered by the younger sediments. The enormous granite batholith that underlies many thousands of square miles of the Sierra Nevada consists of a large number of small intrusions, such as this one, all packed together to make a continuous mass. Apparently this body of granite between Roseville and Auburn was a single bubble of magma that rose alone into the continental crust a few miles away from most of the others.

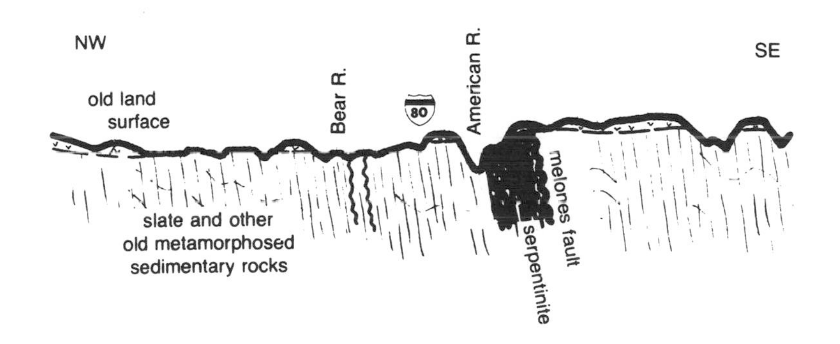

Section across Interstate 80 between Auburn and Emigrant Gap. Streams have cut deep canyons into the old subdued land surface leaving remnants of old gravels and volcanics on drainage divides.

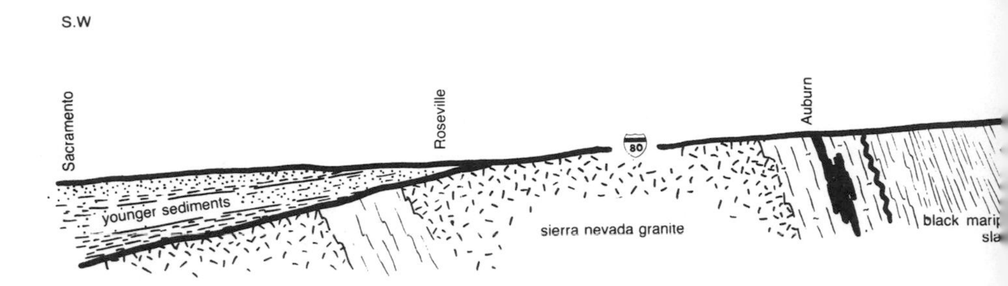

Between Auburn and Emigrant Gap the highway follows a ridge between the valleys of the Bear River, north of the road, and the North Fork of the American River to the south. The crest of the ridge is a remnant of the softly-rolling landscape that existed in the Sierra Nevada before it was tilted and uplifted to become a mountain range. Streams have since bitten canyons deep into the bedrock on either side leaving the old land surface surviving almost untouched as a gently rolling upland on the drainage divide.

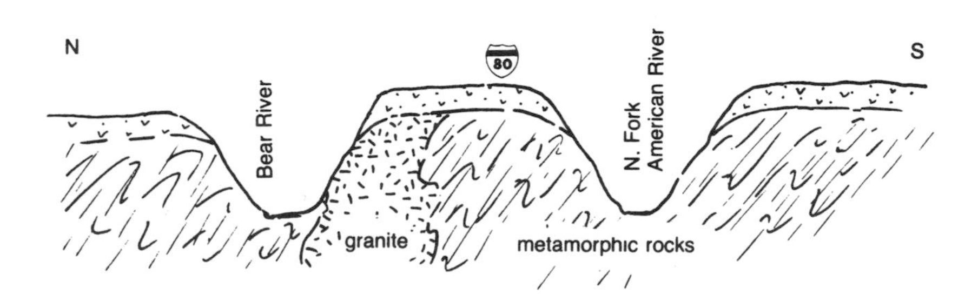

Sketch showing how I-80 follows a remnant of the old sierran land surface perched on the divide between deep canyons eroded by the Bear and American river systems.

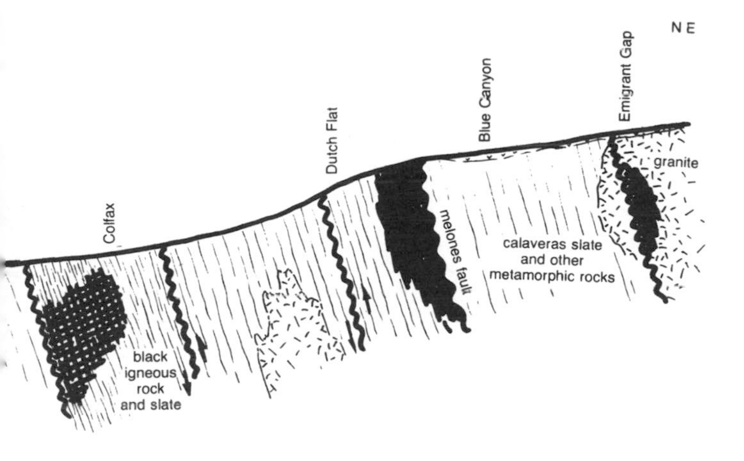

Section along the line of I-80. Great Valley sediments lap onto the complex rocks of the Sierra Nevada west slope.

Very little solid bedrock is visible along this stretch of road. Stream sediments, volcanic materials and a deeply-obscuring mantle of red soil blanket the older landscape developed before the range was uplifted. Coarse stream gravels extensively exposed in spectacular roadcuts near Dutch Flat and Gold Run are the famous "auriferous gravels" eagerly sought by miners of the last century. They contain placer deposits of gold concentrated near the base of the gravels where they lie on bedrock. These are the channels of streams that flowed tens of millions of years ago during the long period of geologic quiet when the Sierra Nevada was a humid lowland crossed by large rivers. Volcanic activity and regional uplift during the last 10 or 15 million years has caused abandonment of old stream courses as modern rivers cut deep new canyons into the old landscape.

Roadcuts in gold-bearing gravels near Dutch Flat. The deposits are reddened by prolonged weathering.

Bedrock in this part of the Sierra Nevada consists almost entirely of old sedimentary rocks originally deposited in the ocean and then mashed against the edge of the continent about 200 million years ago. Intense deformation and prolonged heating have changed the original sediments almost beyond recognition. Now they are slabby rocks that break easily into slaty plates often glistening with flakes of mica. They come in a variety of colors ranging from black to light gray through various shades of tan and even green. But whatever their color these slaty rocks usually look as though they might split into good roofing shingles or stepping stones.

Close view of gold-bearing gravels in roadcut beside I-80 near Dutch Flat. The pebbles are deeply weathered.

EMIGRANT GAP — RENO
(57 miles)

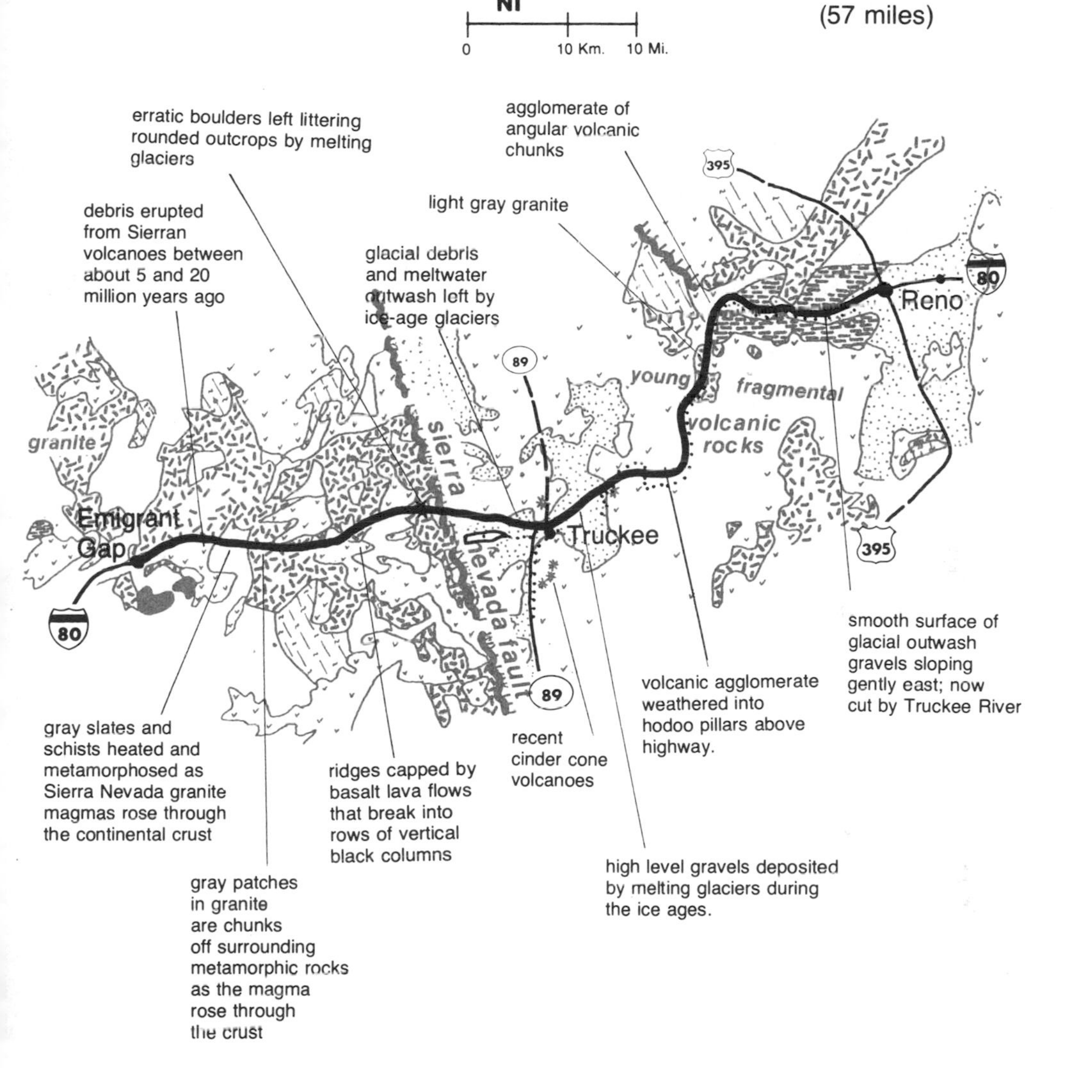

EMIGRANT GAP — RENO

Most of the rock exposed along this stretch of road is granite — massive knobs of smooth gray rock seamed by fractures and littered with boulders rounded by weathering into softly bulging forms. Granites make bold outcrops that invite hikers to seek hand and footholds in the fractures opened into clefts and crevices by the prying processes of weathering. Other rocks along this road are less conspicuous: outcrops of the old sedimentary rocks that had been crushed and metamorphosed into slates before the granites were intruded, and volcanic materials erupted before the Sierra Nevada was uplifted.

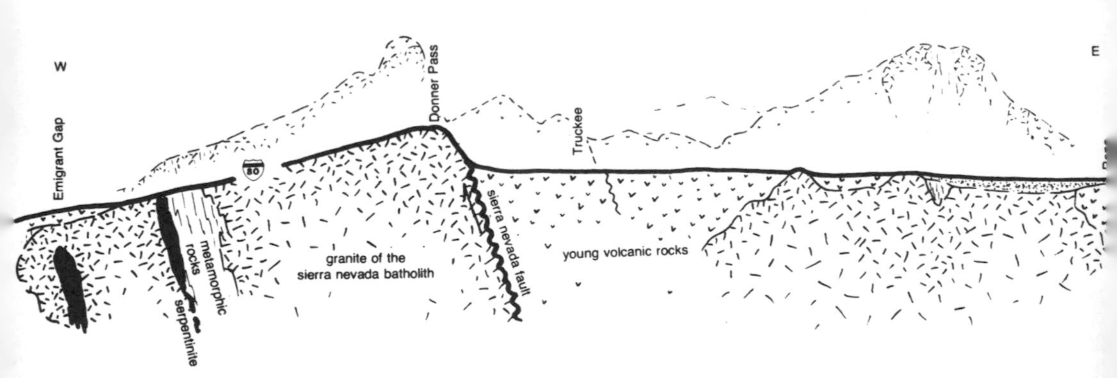

Section along I-80 between Emigrant Gap and Reno. The young volcanic rocks were erupted during the same time that crustal movements were heaving the Sierra Nevada block upward.

Glaciers rasped these high reaches of the Sierra Nevada during the ice ages and left their distinctive mark in peaks bitten away to ragged masses of roughly quarried rock that embrace sparkling little lakes in their hidden recesses. Trees now green these regions so lately inhabited by groaning rivers of ice and they soften, but do not completely hide, the harsh bleakness of the rocky landscape left when the ice melted a few thousand years ago.

Donner Pass is at the crest of the Sierra Nevada almost, but not quite, overlooking the abrupt descent into arid lowlands of the Nevada desert to the east. They have been dry ever since faulting raised the Sierra Nevada into the path of moisture bearing winds from the Pacific several million years ago. The road immediately east of Donner Pass crosses the steep escarpment of the fault surface that defines the eastern front of the Sierra Nevada.

Granite near Donner Pass dappled with dark spots, the cooked remnants of chunks of older rock caught in the molten magma as it rose through the continental crust.

Granite is the dominant rock near Donner Pass, ragged peaks of knobby gray rock rising above valleys scraped smooth by the slow grinding of ice-age glaciers. Those rivers of ice flowed eastward down the front of the Sierra Nevada and onto the floor of the valley below where they dumped their burden of eroded rock. The highway crosses bouldery deposits of glacial material all the way from the west end of Donner Lake to Truckee. In places, buried beneath the glacial deposits along the road, and forming the hills that rise above them on both sides of the road, are volcanic rocks that were erupted onto the valley floor after it was let down at the same time that the Sierra Nevada rose. Most of the rocks exposed east of Truckee all the way to Reno are also volcanics — all erupted within the last 10 or 15 million years.

Between Truckee and Reno the highway follows the valley of the Truckee River which drains water from Lake Tahoe into desert valleys north of Reno where the entire river disappears by evaporation and soaking into the ground. Most streams that flow into deserts disappear in this way before they get very far and the sediment they carry is deposited where the stream ceases to flow. So the flat floors of desert valleys tend to fill as they accumulate sediments eroded from nearby mountains.

Hoodoo pillars eroded into thick deposits of volcanic agglomerate between Truckee and Reno.

50

u.s. 50
sacramento — carson city

Highway 50 crosses the flat floor of the Sacramento Valley and the gently-rolling lower foothills of the Sierra Nevada between Sacramento and Placerville. A few miles east of Placerville it meets the South Fork of the American River which it follows all the way to the crest of the range at Echo Summit, almost to the Nevada line. The river has bitten its valley deep into the bedrock core of the Sierra Nevada since the range began its long uplift along faults at its eastern margin several million years ago. Rocks exposed along its canyon show a cross-section through the profound bedrock architecture of the range. They provide a much different impression from that conveyed by roadcuts along Interstate 80 which follows a parallel route high along the drainage divide just a few miles to the north — across remnants of the landscape that existed before the range was lifted. East of Echo Summit the road works its way down the steep eastern face of the Sierra Nevada created by faulting during the last few million years.

SACRAMENTO — PLACERVILLE
(45 miles)

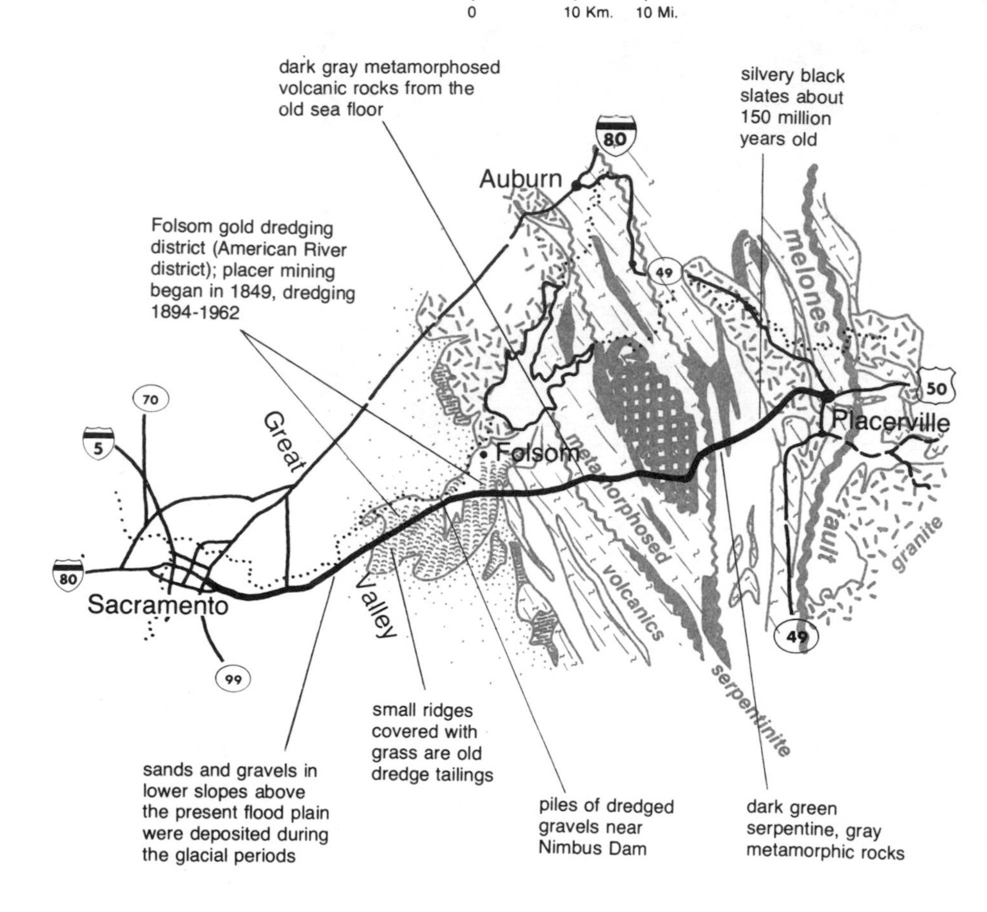

SACRAMENTO — PLACERVILLE

Highway 50 follows the broad valley of the American River east from Sacramento to where it leaves the Sacramento Valley and crosses into the foothills of the Sierra Nevada just south of Folsom Reservoir, a distance of about 20 miles. Floodplain deposits underlie the road and the low valley walls are cut into deposits of sediment washed from the Sierra Nevada onto the flat valley floor within the last few million years.

The eastern half of the route between Sacramento and the point where the road enters the rolling sierra foothills crosses old placer mine tailings. Nearly ten miles of road cross the middle of an ugliness of mounded heaps of gravel left, without apology, by people who ruthlessly overturned the entire floodplain in their frantic search for gold. These monuments to greed may survive to become an enduring relic of our civilization.

Bouldery dredge tailings in the floodplain of the American River.

It is astonishing to see the coarseness of the gravels in placer mine tailings; they generally contain pebbles much larger than any that lie on undisturbed gravel bars in the same stream. Such gravels lurk in the deepest parts of stream channels, dozens of feet below the surface, where they remain undisturbed by all except the greatest floods. Very occasionally, perhaps once or twice in a thousand years, stream valleys are scoured by floods powerful enough to move the gravels in the bottoms of the channels. During these times when the entire stream bed is moving any gold that may be around sinks to the bottom of the channel. To recover the gold, placer miners must sift through the deepest and coarsest gravels so they completely overturn the floodplain. Because their debris piles contain pebbles too coarse for the stream to move during any but the greatest floods, thousands of years must elapse before nature can repair the damage.

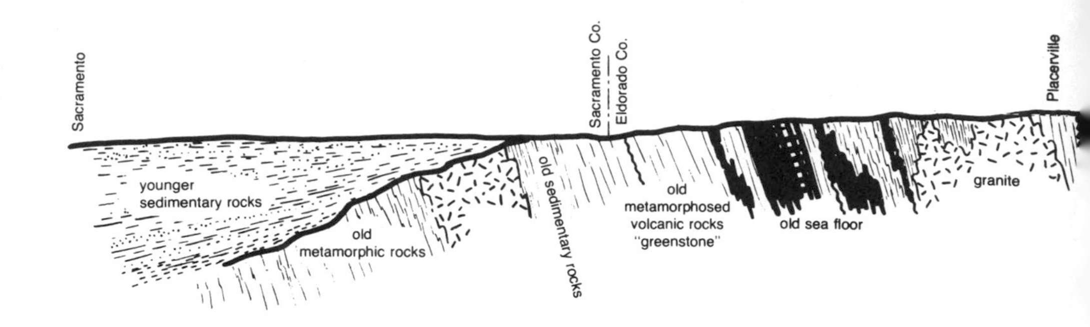

Section between Sacramento and Placerville. Great Valley sediments lap onto the much older rocks of the Sierra Nevada.

Early placer miners worked by hand digging the gravel from the bottom of a shaft and trundling it to a small sluice box in wheelbarrows. Their tailings, still visible in many remote corners of the Sierra, are hummocky expanses of small heaps of gravel. Mechanization came to placer mining in the form of the dredge which floats in a pond of its own making and scoops gravel from the bottom of the channel with an endless chain of buckets, washes it, and dumps the tailings by conveyor belt at the back of the pond. Dredge tailings are generally in rows of heaps suggesting giant furrows. Gold dredging began in California during the latter part of the

An old gold dredge slowly moldering in its last bit of pond. This one operated in California for many years before moving to Montana for its last stint.

nineteenth century and a few dredges continued their depredations until the early sixties when they were finally stopped by economic conditions. All of the tailings visible from U.S. 50 were left by dredges.

Between the point where it leaves the ravaged floodplain of the American River and Placerville, U.S. 50 crosses a broad belt of old sedimentary rocks metamorphosed into slates and schists by intense deformation and recrystallization. Outcrops are few along this part of the route that cuts across the old, deeply weathered terrain eroded before the Sierra Nevada was uplifted to become a range of mountains. Rocks in this area were once muddy sediments until they were crushed against the edge of the continent during the early stages of the development of Califonia. Large masses of molten granite magma intruded them later, while the Franciscan rocks were being jammed into a marginal trench. Gold-bearing quartz veins, formed in fractures in this area while the granite magmas were intruding.

PLACERVILLE — CARSON CITY
(89 miles)

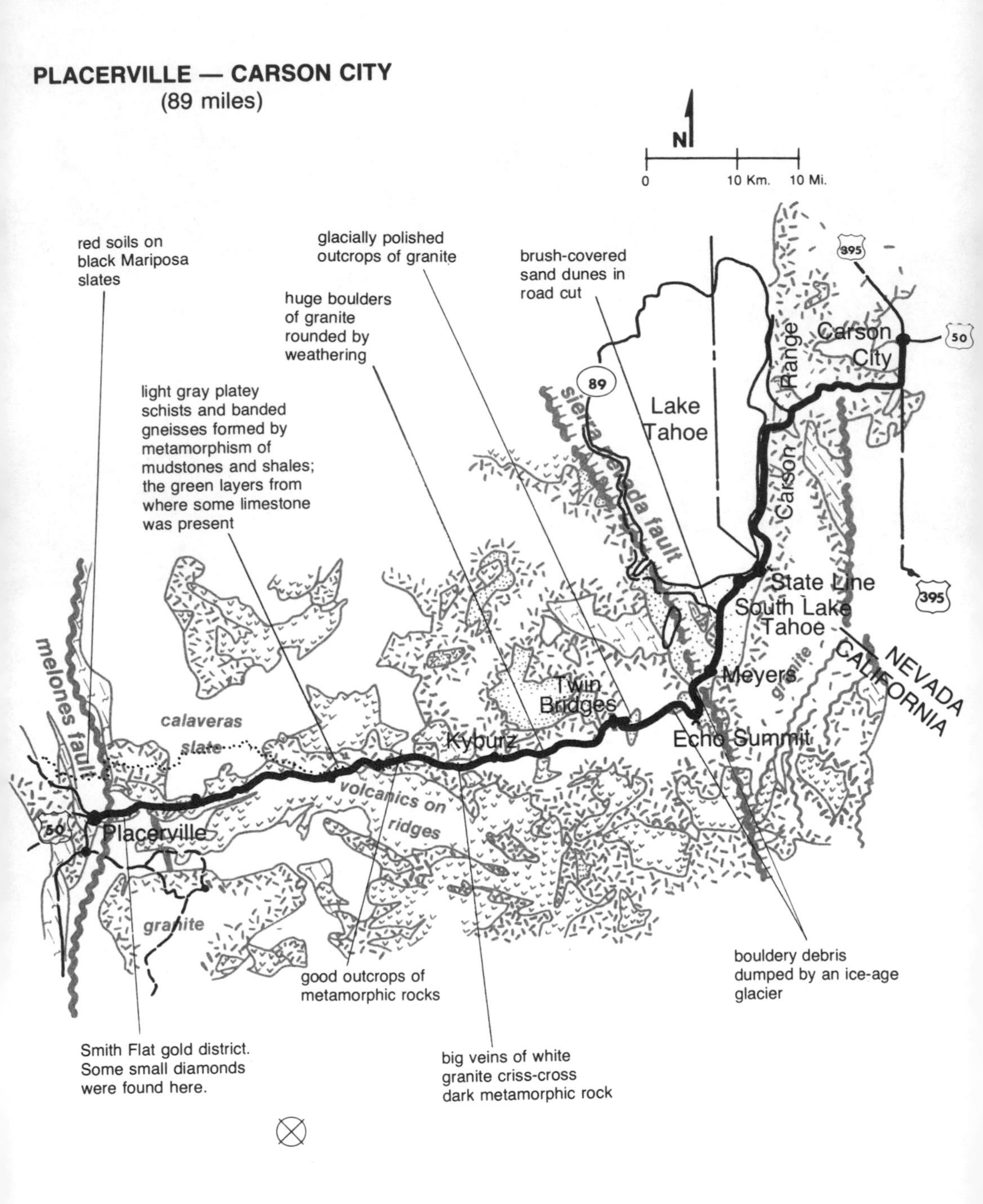

Streaky gneiss injected by a dike of granite.

Slates and schists make slabby outcrops.

PLACERVILLE — CARSON CITY

Most of the route between Placerville and Riverton crosses volcanic rocks erupted onto the old pre-uplift landscape of the Sierra Nevada during the Pliocene Period, 5 or 10 million years ago. These are mostly deposits of light-colored volcanic ash, soft rocks that weather easily and rarely make prominent outcrops. They are preserved here, perched on a drainage divide, as uneroded remnants of the old landscape now being deeply carved by the modern streams.

Between the small communities of Pollock Pines and Riverton the road descends into the valley of the South Fork of the American River at a place where it is eroded through the oldest sequence of sedimentary rocks exposed on the western slopes of the Sierra Nevada. These rocks were first deposited in the ocean between 200 and 300 million years ago, long before the sea floor began descending beneath this part of the continent, sweeping its veneer of sediments into ranges of coastal mountains. They form a broad belt along the higher foothills of the Sierra Nevada but very little is exposed along U.S. 50. Like the younger sedimentary rocks to the west, which they resemble, these rocks contain numerous veins of gold-bearing quartz.

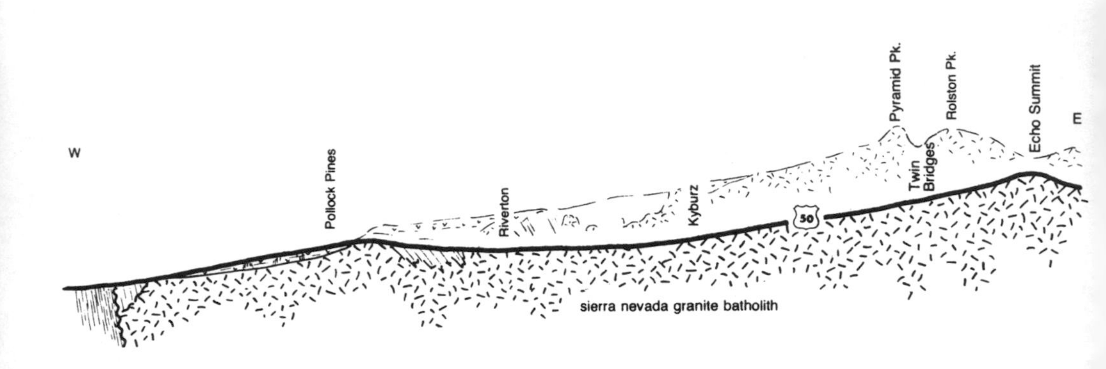

Section along the line of U.S. 50 between Placerville and the crest of the Sierra Nevada. Most of the route crosses parts of the enormous Sierra Nevada granite batholith.

Almost all of the bedrock between Riverton and Echo Summit is granite, part of the great Sierra Nevada batholith. Bold outcrops of massive, gray rock rounded by weathering into bulging forms that often suggest hugely swollen pillows. Granites generally weather into friendly landscapes that invite wandering among the rounded outcrops and the trees. Close examination of the rock reveals its coarsely-crystalline texture composed of rectangular grains of milky-white or salmon-pink feldspar set among glassy grains of quartz, usually irregular in form. Glossy-black crystals, flakes of biotite mica or stubby needles of hornblende, pepper most granites.

Glaciers gnawed the crest of the Sierra Nevada during the great ice ages of the recent geologic past, as recently as 10,000 years ago. They left their erosional signature in craggy peaks, carved into bare rock, with numerous small lakes huddling in rocky basins between them. Echo Lake at the summit of the pass is one of these.

Between Echo Summit and Meyers, a distance of only a few miles, U.S. 50 crosses the abrupt fault escarpment that defines the steep eastern face of the Sierra Nevada. This fault is still moving, to the accompaniment of occasional earthquakes, and in many places there is good evidence that the Sierra Nevada is still getting higher. All of the bedrock along the way is granite but much of it is buried beneath a thick plaster of bouldery glacial debris.

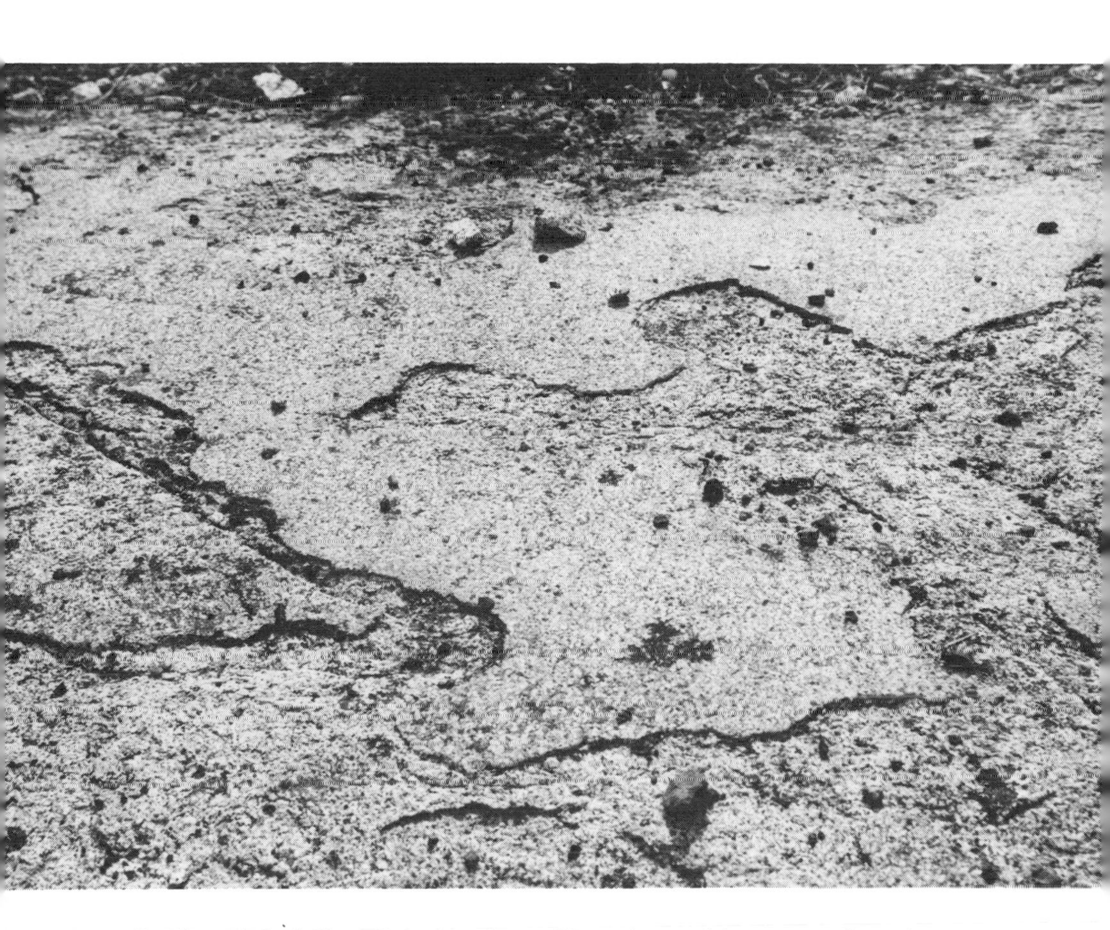

Patches of glacially-polished surface still survive on this outcrop of granite near Twin Bridges.

Boulders of granite in glacial deposits beside U.S. 50 west of Echo Summit.

Most of the route between Meyers and Carson City is across the floor of a down-dropped block of the continental crust. Part of the distance is along the east shore of Lake Tahoe, the foremost gem of the Sierra Nevada, now in danger of being converted into a stinking cesspool by its admirers of whom there are evidently far too many. The Carson Range along the east side of Lake Tahoe, is composed of the same granite that forms the crest of the Sierra Nevada.

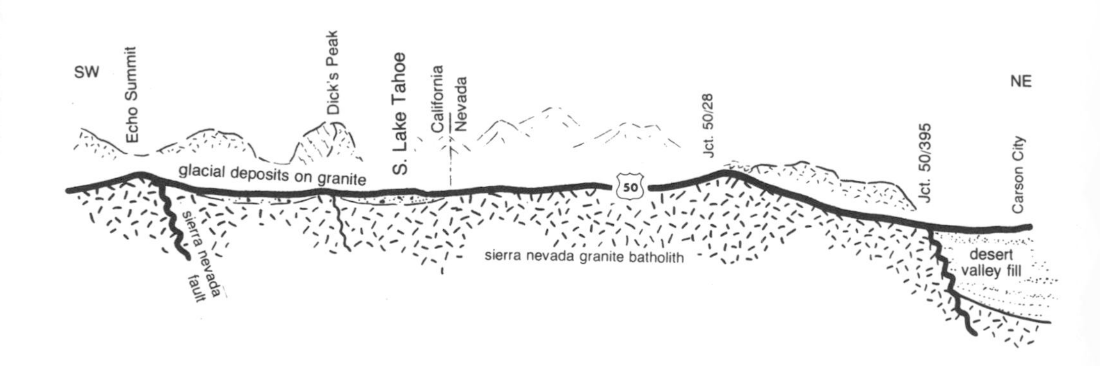

Section along U.S. 50 from Echo Summit to South Lake Tahoe and Carson City.

WEST SHORE: SOUTH LAKE TAHOE — TAHOE CITY
(27 miles)
EAST SHORE: SOUTH LAKE TAHOE — TAHOE CITY
(44 miles)

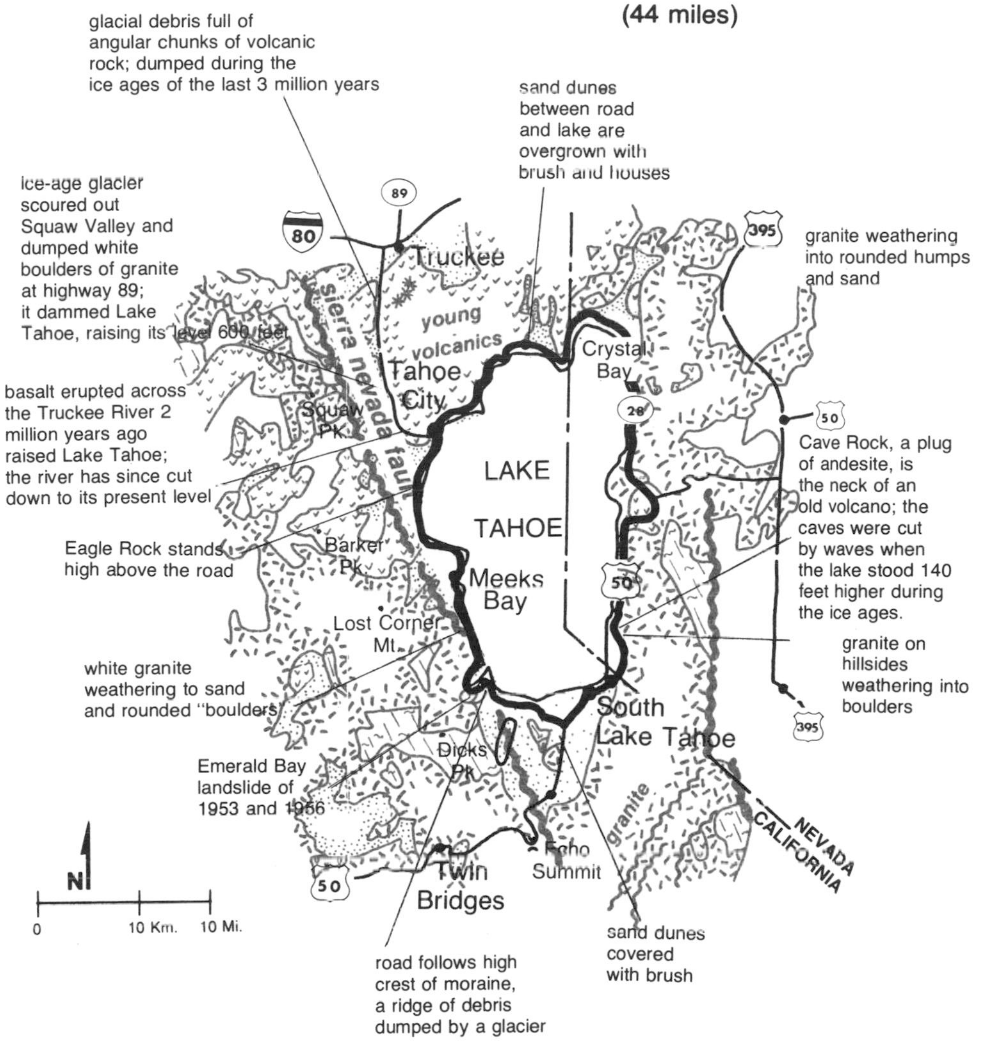

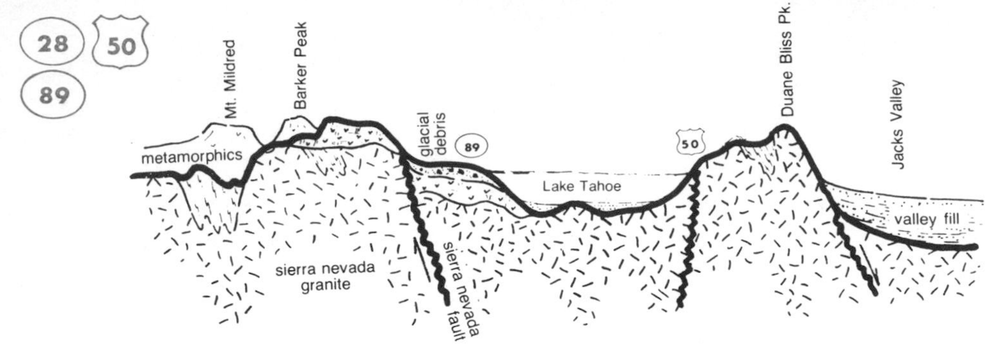

Tahoe floods a dropped crustal block.

lake tahoe — gem of the sierra

Like all large lakes, Tahoe exists because of a series of geologic accidents. Erosional processes tend naturally to shape the landscape so that surface waters drain away to the sea. Lakes form only where something has happened to interrupt drainage and they last only as long as it takes the slow processes of erosion to restore it. Streams draining into and out of lakes fill their basins with sediment and cut their outlets lower to restore drainage.

Lake Tahoe fills a valley created within the last several million years as the continental crust in this region was breaking along faults into large blocks. Some of the blocks, such as the Sierra Nevada and Carson Range on opposite sides of Lake Tahoe, rose to become mountains while others, such as the one beneath Lake Tahoe, sank to become valleys. Eruptive activity that accompanied the block faulting built a pile of volcanic rocks completely across the valley to create the natural dam impounding Lake Tahoe. Interstate 80 crosses the volcanic dam between Truckee and Reno but good exposures are few because much of that route is well plastered with younger glacial deposits.

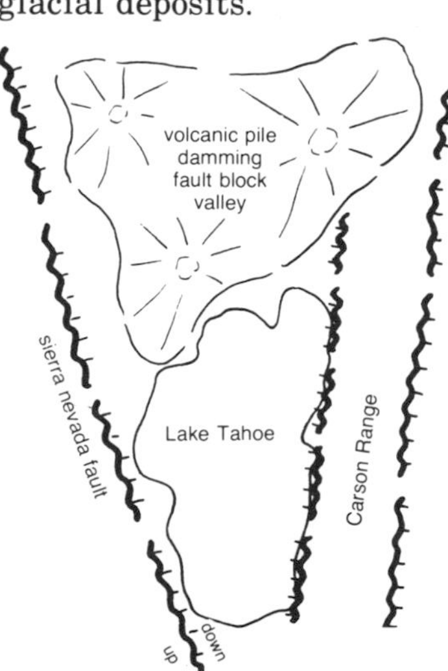

A pile of volcanic material impounds Lake Tahoe.

Drainage from a large area of the wet eastern crest of the Sierra Nevada empties into Lake Tahoe keeping it full of sparkling clear water. Overflow empties into the Truckee River, a modest little stream doing its best to drain Lake Tahoe by cutting the outlet channel deeper. But that is thousands of years in the future; now the Truckee River runs brightly through the mountains to Reno and then withers away to its wretched destiny in the parched Nevada desert around Pyramid Lake. There is not enough water to run the stream any farther through the desert.

During the last ice age, two of the many enormous glaciers that formed in the Sierra Nevada ground their way down Squaw Valley and Pole Creek to the Truckee River where they dammed the outlet of Lake Tahoe, raising the water as much as 600 feet. The ice dam floated several times, releasing catastrophic floods which carried huge boulders downstream past Truckee. When the ice melted 10,000 years ago, the lake returned to its formed level. These adventures left their mark in abandoned beaches and high benches of near-shore deposits that survive as lasting souvenirs of the ice ages. The sedimentary benches make large terraces, generally heavily forested, that rise above long stretches of the lake shore looking like a giant stair step and tread leading up to the mountains. Their flat tops are irresistably tempting to developers but construction of any kind is likely to be hazardous because the soft muds within the benches are very weak and likely to slide.

Landslides have been a problem in the Lake Tahoe area. A very large mass began to move above Highway 89 at the head of Emerald Bay in 1953 and then came down to obliterate several hundred feet of road during a long rainy spell two years later. Weak bedrock was the basic cause of this slide — granite that had

Emerald Bay is enclosed by the arms of a moraine recording the margins of an ice-age glacier that once stood here.

The Emerald Bay slide.

been thoroughly broken in a fault zone and then softened still more by weathering along the fractures. There have been a number of other landslides in the Tahoe area and the problem of avoiding future ones will become serious if the present rate of development continues.

Of course the major threat Lake Tahoe faces is the prospect that people may become so numerous and careless that they will pollute its unbelievably clear water. Tahoe is especially vulnerable to pollution because its drainage basin is very small in proportion to its volume so the amount of water entering and leaving the lake each year is also very small. If the lake were somehow drained, it would take about 600 years to naturally refill its basin to the present level. Therefore, any pollutants dumped into the lake will remain there for hundreds of years before they can be naturally flushed. Even thoroughly treated sewage effluent is a problem because it contains fertilizer nutrients that stimulate growth of algae which have already become much more abundant that they were a few years ago. The only solution seems to be export of sewage effluent from the drainage basin, strict regulation of septic tanks, and extreme care in land management throughout the entire drainage basin. The alternative is to watch Lake Tahoe slowly but permanently turn into a slimy green soup of floating algae.

u.s. 50, nevada 28

south lake tahoe — tahoe city
east shore

The road closely follows the east shore of the lake where it washes against the granite flank of the Carson Range all the way from South Lake Tahoe to Crystal Bay. There are magnificent views west across the lake to the alpine valleys and glacially carved crags of the high Sierra Nevada.

Bedrock along the east shore is almost entirely sierran granite except for a small patch of young volcanic rocks at Glenbrook Bay and an eroded volcanic neck about 7 miles north of Stateline. Caye Rock is part of that volcanic neck named for its caves that were eroded by waves during the ice age when the lake level stood much higher because the outlet was dammed by a glacier between Tahoe City and Truckee.

Bedrock along the route between Crystal Bay and Tahoe City is entirely young volcanic rocks erupted within the last 10 million years. These rocks are the natural dam that impounds Lake Tahoe in its fault-rimmed bedrock basin. Along the flat north sides of Crystal and Agate Bays the volcanic bedrock is completely buried beneath lake sediments and sand dunes deposited during the very recent geologic past.

Weathered granite along the east shore.

california 89

south lake tahoe — tahoe city
west shore

The road hugs the western shoreline of the lake where it laps the steep eastern slope of the Sierra Nevada. The view east across the lake is of the Carson Range, granite mountains that were hardly touched by ice-age glaciation. Their peaks are full and rounded, not craggy and jagged like those of the glaciated high Sierra which caught most of the snow.

Bedrock along the southern half of the west shore is mostly granite but much of it is well buried beneath large glacial moraines composed mostly of granite boulders. These were deposited by rivers of ice that descended from the crest of the range into the waters of the lake. Icebergs must have broken off them and drifted in the lake during the last ice age. Granite in the vicinity of D.L. Bliss State Park escaped glaciation, so preserving the deep weathering and thick sandy soils developed before the ice ages. A short but steep valley glacier descending from the west scoured out the basin of Emerald Bay. Melting where it reached Lake Tahoe, it dumped its load of dirt and boulders to form glacial moraines that define the bay.

Granites usually weather to rubbly soils which have a very open texture.

This jumble of rough boulders beside California 89 is a volcanic agglomerate probably deposited by a mudflow.

TAHOE CITY — BLAIRSDEN
(63 miles)

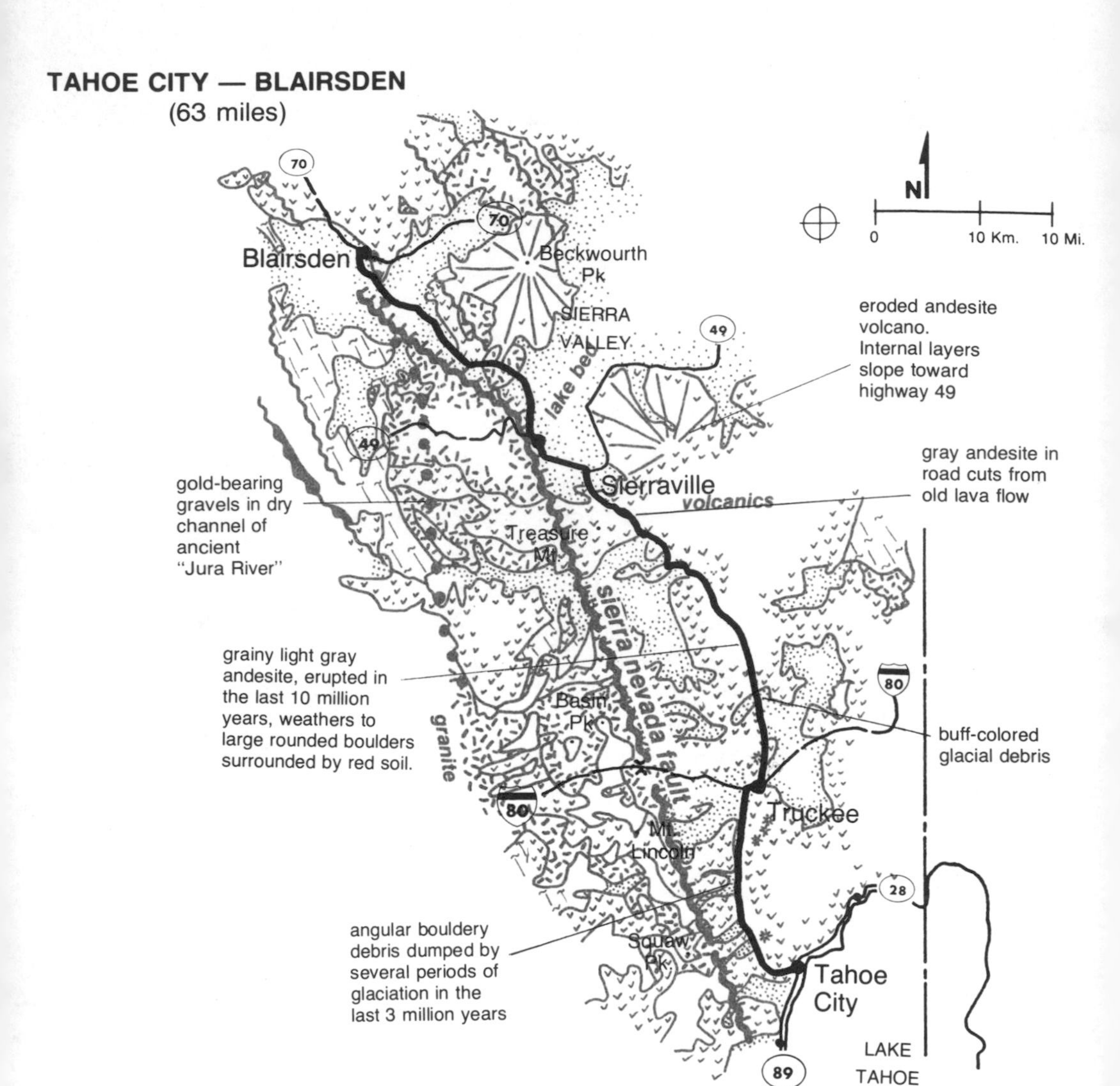

Eagle Rock, the neck of an eroded basalt volcano, stands about 4 miles south of Tahoe City.

TAHOE CITY — BLAIRSDEN

Between Tahoe City and Blairsden Highway 89 follows a route within a few miles of the base of the steep fault escarpment that is the eastern face of the Sierra Nevada. The high peaks of the Sierra outlined on the western skyline are carved from the same granites that lie deeply buried beneath the younger volcanic rocks along the highway.

All of the bedrock exposed along the road between Tahoe City and Blairsden is volcanic, all of it erupted within the last 10 million years — during the period in which this crustal block was subsiding along faults and the Sierra Nevada rising. As usual with volcanic rocks, these come in disguises as various as the different ways volcanoes can erupt. There are a few basalt lava flows, easy to recognize by their absolute blackness, and a few beds of white volcanic ash blown violently from bodies of molten granitic magma. Many of the rocks are andesites, intermediate between basalt and rhyolite in both chemical composition and color. Some of the andesites are lava flows which are recognizeable as solid rocks containing a few crystals scattered through a fine-grained brown matrix. Most are messy-looking agglomerates made of ash and various-sized chunks all mixed together.

Some of the older volcanoes have been around long enough to be carved by erosion but the younger ones still have their original form. This complex landscape does not give the impression of being the floor of a vast bedrock basin formed by faulting; that is a story told by the rocks.

Between Tahoe City and Truckee the volcanic rocks are mostly covered by younger glacial deposits. In some exposures the two look very much alike. The volcanic rocks contain numerous rather messy mudflow deposits which the glaciers scooped up and converted into rather messy glacial deposits. Both are full of angular chunks of lava in a variety of pastel and reddish colors but the glacial deposits also contain speckled chunks of gray granite eroded from the crest of the Sierra Nevada.

Painstaking and detailed study of glacial deposits in this area has shown that they record at least five major episodes of glaciation separated by periods in which the glaciers were melted. Here, as elsewhere, there was not one ice age but a series of them during the last three million or so years. Probably there will be more in the future.

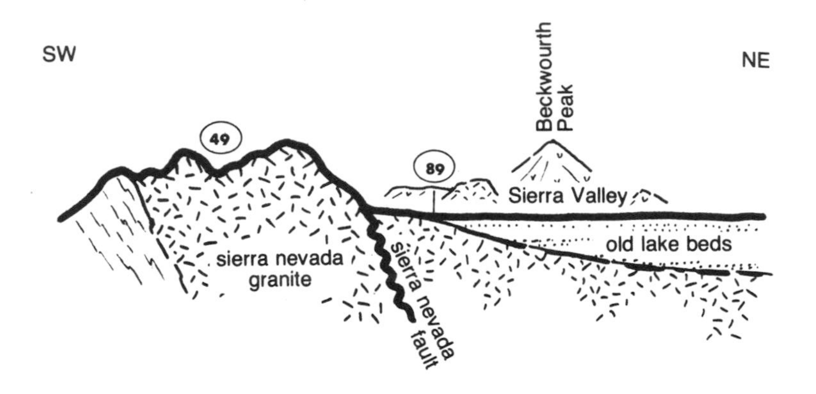

Section across the steep eastern face of the Sierra Nevada. Most of the hills in the Sierra Valley are volcanoes.

Between Sierraville and Blairsden, the road crosses the west side of the Sierra Valley, an area covered by lake deposits instead of volcanics. Apparently this basin was dammed by volcanic deposits and filled with water until the outlet stream eroded its valley deep enough to restore drainage. Flowing artesian wells in the Sierra Valley produce water trapped under pressure beneath the fine-grained and impermeable lake sediments that surface the flat floor of the valley.

california 49

placerville — vinton — gold country

The mother lode highway winds among the western foothills of the Sierra Nevada passing through many of the historic mining communities established during the great gold rush. A large proportion of all the gold ever mined in California was recovered within sight of this narrow road from a region that was once the busiest part of California. The mines are closed now and the mother lode country seems quiet and remote, as though nothing had ever happened.

It was not lack of gold that closed the mines as much as it was the economics of recovering it. Because its principal use is as a medium of international monetary exchange, the price of gold has been fixed by law for many years with the government as the only legal customer. While the cost of mining rose over the years, especially the cost of labor, the price of gold remained fixed, catching the mining companies in a squeeze that eventually closed all their operations. Plenty of gold still remains in the Sierra foothills and the recent sharp rise in its domestic price is bringing a resumption of mining.

Americans will be happier if economic conditions never favor large-scale placer gold mining. Unlike most other industries, gold mining thrives on hard times — on unemployment and tight money. When people find it impossible to earn a living in other ways, they attempt to dig money out of stream gravels. The last big surge in gold production in California came during the depression of the 1930's when output briefly reached levels approaching those of the mid-nineteenth century. But that income was pathetically small when averaged over the many thousands of people who were then engaged in mining. Now that the passage of many years has enabled the ravaged streams to clean their channels and a growth of trees to start in the yawning earth wounds opened by hydraulic mining, it is easy to cast a romantic spell over the memory of placer gold mining in California. It is hard to remember in the stillness of a ghost town that it was the site of a squalid industry driven by greed and economic desperation and producing a commodity useful mainly for storage at Fort Knox.

Placer and hydraulic mining dumps enormous quantities of mud, sand and gravel into nearby streams choking their channels with deposits that build the stream bed higher. This gives the stream a steeper gradient enabling it to flow faster and carry the extra sediment load — otherwise it would be permanently choked. When gold mining in the Sierra Nevada and Klamath Mountains during the last century caused the streams to build up their beds by deposition, it also caused them to flood vast acreages of productive farmland in the Sacramento Valley. Legal disputes followed, culminating in the Sawyer decision of 1884 forbidding miners to dump debris in streams thus effectively stopping hydraulic gold mining in California. The environmental costs were too much even for that ruthless period. The expense of building debris dams to contain tailings is so great that few deposits are now profitable to mine.

High gravel banks still remain in the channels of many Sierra Nevada streams, souvenirs of hydraulic mining as are the old mine pits several of which are near Highway 49. Curiously, the old hydraulic pits, once thought to be hopeless, are now beginning to be forested. Evidently the bare gravels exposed by mining are a fertile seedbed for certain kinds of trees.

Quartz veins were worked underground by sinking shafts and then mining them out and hauling the rock to the surface where it was crushed and milled to free the gold. Although they were nasty places to work, these mines did not deface the countryside and make nuisances of themselves on nearly the same scale as the placer and hydraulic workings. Many of them operated for over a century before economic pressures finally closed the last of them during the decade following the second world war. A few of the underground operations in the "mother lode" were very large and unbelievably productive. One of the northern mines at Grass Valley reached a vertical depth of slightly more than a mile, had more than 200 miles of underground workings, and is reported to have produced more than $120,000,000 in gold bullion.

South of Grass Valley, Highway 49 trends generally north-south, cutting across the topographic "grain" of the country by

Head frame of the "Rowe Shaft" to the Empire Mine at Grass Valley.

rising over drainage divides and winding in and out of the major valleys. Between stream valleys the road crosses remnants of the old gently-rolling landscape that existed before the region became a range of mountains. Deep red soils mantle this old landscape making bedrock outcrops scarce. Erosion has entrenched the major stream valleys since the range was uplifted and it is usually possible to see good outcrops there because the deep soil is missing. A few miles north of Grass Valley the road turns east and follows the canyon of the Yuba River into the heart of the Sierra Nevada and then on into the arid valley and range country to the east.

PLACERVILLE — GRASS VALLEY
(50 miles)

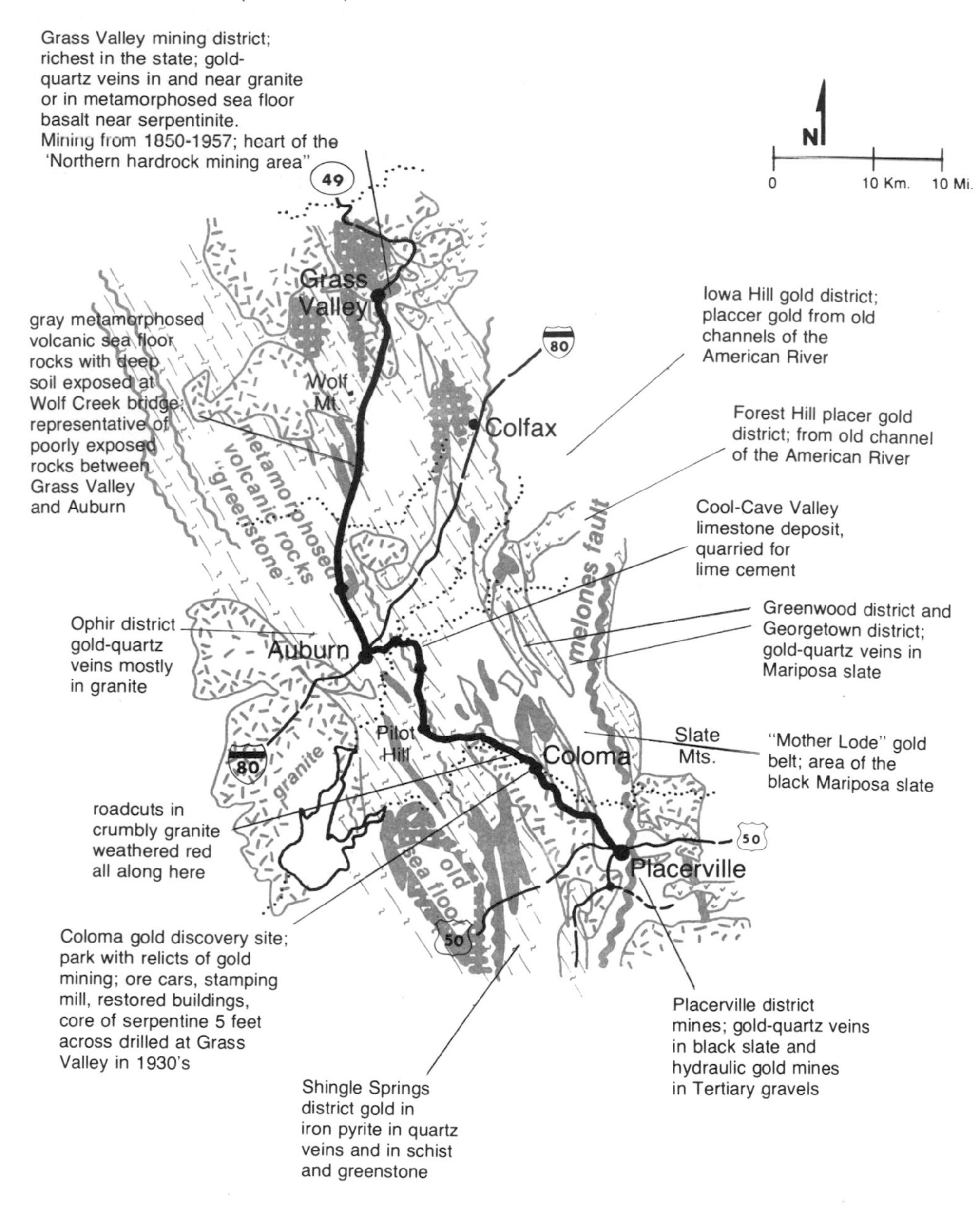

PLACERVILLE — GRASS VALLEY

Placerville is on a remnant of the old landscape of softly rounded hills underlain by deep red soils that existed before uplift of the Sierra Nevada, and so is Auburn. Between them the highway winds through deep and rugged valleys eroded by the American River and its tributaries since uplift of the range began a few million years ago.

Bedrock near Placerville is almost entirely sedimentary rock, mostly layers of muddy sandstone, originally deposited on the seafloor and then jammed into the range of coastal mountains about 150 million years ago. Bedrock near Auburn has essentially the same history but differs in its content of large quantities of volcanic material — easy to recognize as distinctly green layered rock. There is an intrusion of granite between Auburn and Placerville which the road crosses for a distance of about 11 miles where it follows the South Fork of the American River, in the vicinity of Lotus and Coloma.

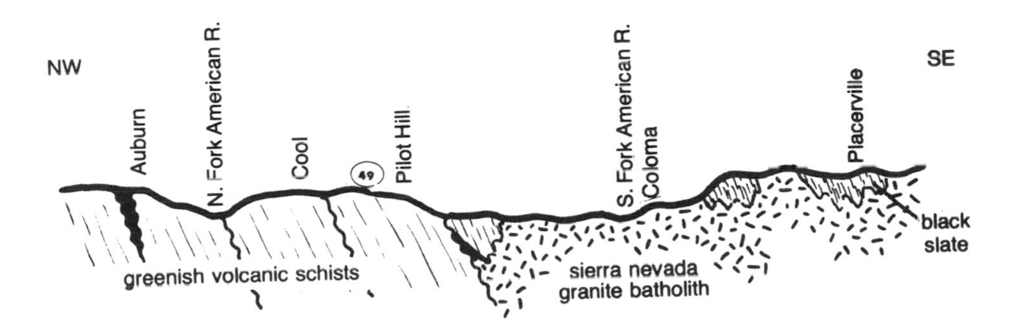

Section along the line of California 49 as it passes over the crest of the Sierra Nevada between Downieville and Vinton.

Gold was first discovered in California at Coloma in gravels along the South Fork of the American River. It was a small deposit of placer gold and not a major find but enough to start a major gold rush. Several other small placer deposits were worked in the neighborhood of Coloma but there were never any large mines. The main trend of the mother lode is a few miles northeast of Highway 49, near Georgetown — another picturesque area that never saw major mining operations.

About a mile and a half north of Cool there are some old quarries in a layer of limestone, a sedimentary rock composed of the mineral calcite — most often derived from seashells. Although it is a very common rock in many areas, relatively little limestone is deposited on the deep sea floor so it is rarely found in the western Sierra Nevada or the Coast Range where the sedimentary rocks are mostly deep sea sediments. Limestone has many important industrial uses, it is a vital ingredient of smelter fluxes and portland cement and is also widely used in the chemical industry.

Most of the scenery between Auburn and Grass Valley borders on being flat because the route crosses extensive remnants of the old land surface. Deeply weathered red soils hide the bedrock almost completely — most of it consists of green to gray volcanic rocks and layers of muddy sediments scraped off the seafloor about 150 million years ago. Several bodies of granite are along the way but the highway manages to miss crossing all except the one just south of Grass Valley. About 9 miles north of Auburn the highway crosses the Bear River which has cut its valley deeply into the old landscape. Good outcrops of green volcanic rocks cut by faults and intruded by small masses of darker green serpentinite add interest to a walk along the river in either direction from the bridge.

Water-driven Pelton wheel powered machinery at the North Star mine in Grass Valley – now an interesting minimg museum.

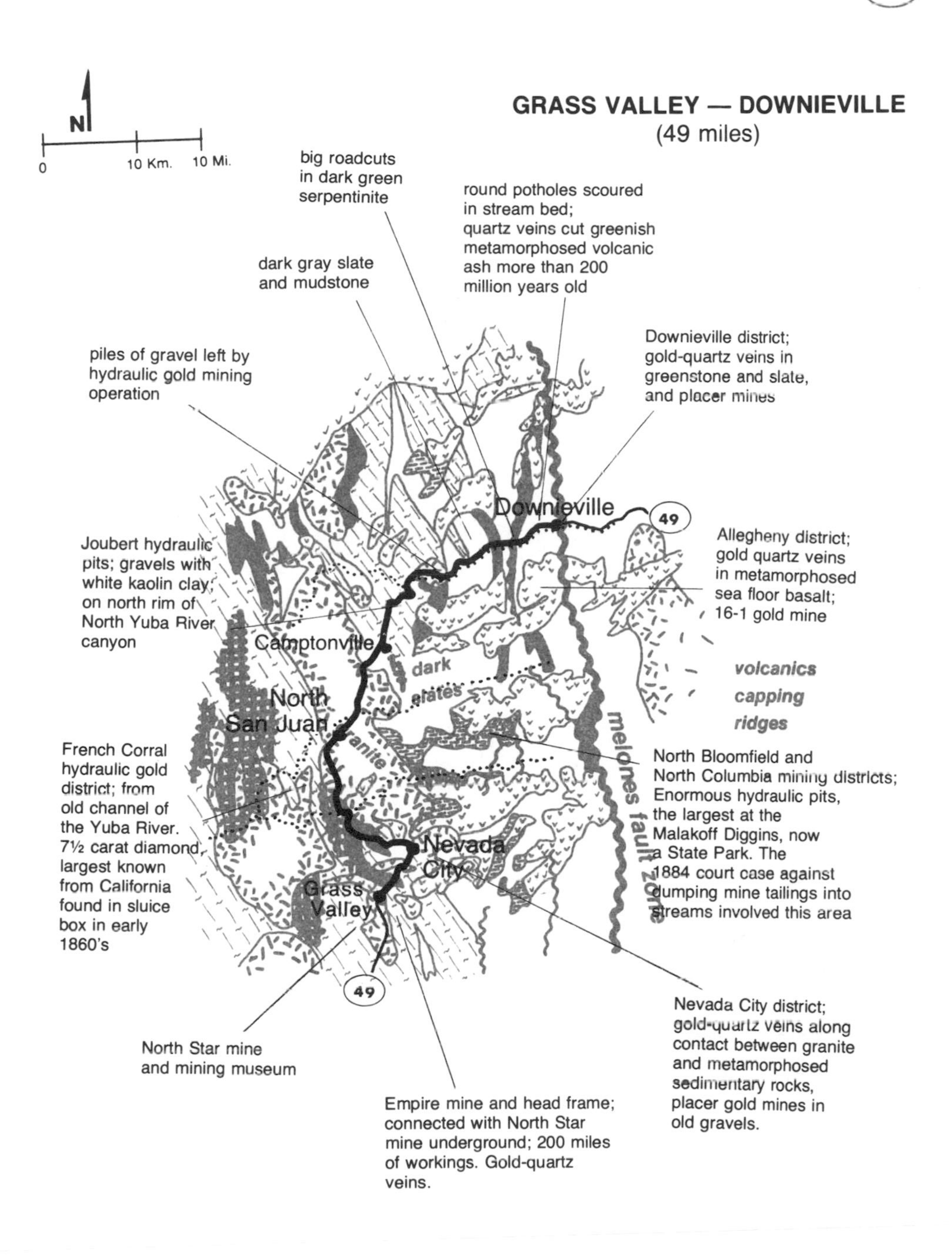
49
N
0
10 Km.
10 Mi.
GRASS VALLEY — DOWNIEVILLE
(49 miles)
big roadcuts in dark green serpentinite
round potholes scoured in stream bed; quartz veins cut greenish metamorphosed volcanic ash more than 200 million years old
dark gray slate and mudstone
piles of gravel left by hydraulic gold mining operation
Downieville district; gold-quartz veins in greenstone and slate, and placer mines
Downieville
49
Allegheny district; gold quartz veins in metamorphosed sea floor basalt; 16-1 gold mine
Joubert hydraulic pits; gravels with white kaolin clay; on north rim of North Yuba River canyon
Camptonville
dark
slates
volcanics capping ridges
North San Juan
granite
melones fault zone
French Corral hydraulic gold district; from old channel of the Yuba River. 7½ carat diamond, largest known from California found in sluice box in early 1860's
North Bloomfield and North Columbia mining districts; Enormous hydraulic pits, the largest at the Malakoff Diggins, now a State Park. The 1884 court case against dumping mine tailings into streams involved this area
Nevada City
Grass Valley
49
North Star mine and mining museum
Nevada City district; gold-quartz veins along contact between granite and metamorphosed sedimentary rocks, placer gold mines in old gravels.
Empire mine and head frame; connected with North Star mine underground; 200 miles of workings. Gold-quartz veins.

GRASS VALLEY — DOWNIEVILLE

Grass Valley and the neighboring town of Nevada City, four miles away, were both major mining camps for nearly a century.

Hundreds of miles of underground workings burrow along the quartz veins beneath these towns. The California Division of Mines and Geology has estimated that the total combined production of these two districts is in excess of 350 million dollars, most of it from the underground quartz mines.

Thick gravel deposits on the western edge of Nevada City were washed by hydraulic miners about a century ago. Old workings, now thickly overgrown, are still visible on both sides of the road which goes right through them. Just over a mile west of Nevada City there are some good roadcuts where the highway crosses onto the border of a large intrusion of granite. Numerous dark patches in the granite are all that remain of chunks of sedimentary rock that got into the molten magma and were very nearly melted themselves before everything solidified. Such inclusions are common in granitic rocks, especially near the margins of intrusions where pieces of the older wall rock would naturally be most abundant.

Thousands of old mine cars like this once trundled along beneath the hills in the "Mother Lode" country. A few of them rust beneath the sky and a lot more underground.

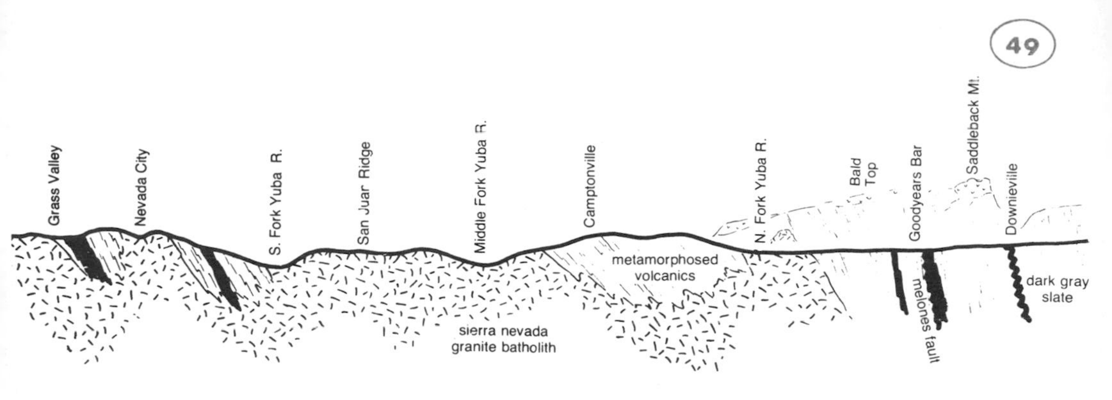

Section along the line of California 49 between Grass Valley and Downieville.

Most of the route between Grass Valley and Camptonville crosses the large granite intrusion that extends to within a mile of both towns. Local outcrops of other kinds of rock, layers of dark sedimentary rocks, green volcanics and some deeply-weathered serpentinite, are older rocks enclosed by the granite. Spectacular outcrops of granite near the bridge over the Yuba River about midway between Nevada City and North San Juan, are forbidding faces of gray rock seamed by widely-spaced fractures and a few veins. Granites normally have widely-spaced fractures that divide the rock into enormous blocks, weighing many tons, which become boulders as their sharp edges and corners are rounded off by weathering.

Very probably it was heat from the large granite intrusion between Grass Valley and Camptonville that moved the circulating hot waters responsible for emplacing the rich gold veins in this area. Big granite intrusions everywhere have a way of spawning gold mining camps along their borders.

An outcrop of black slate.

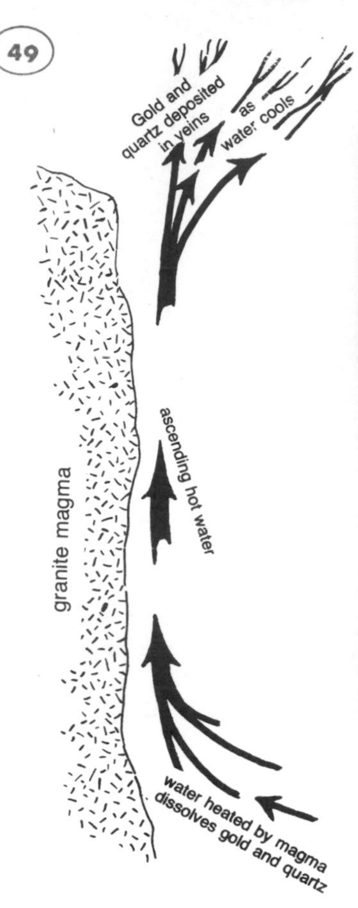

Stretched and broken veins of white quartz deposited in black slate by circulating hot water.

About a half mile south of Camptonville the road leaves the granite intrusion and crosses back onto older sedimentary and volcanic rocks, much deformed and changed by recrystallization, which it follows most of the way to Downieville. These are still the same scrapings off the seafloor that Highway 49 crosses along most of its route north from Placerville.

About five miles north of Camptonville, after passing several more old hydraulic pits, the road winds down into the deep gorge of the North Fork of the Yuba River which it follows all the way east to Downieville and beyond to Yuba Pass. Erosion has cut this deep valley since uplift of the Sierra Nevada began several million years ago. Remnants of the old pre-uplift land surface are perched on drainage divides at the very top of the landscape visible from the road, as much as 1000 feet above the level of the river.

Polished fracture surface – "slickensides" – in dark green serpentinite. Near junction to Goodyear's Bar.

Exposures are good along most of the canyon and include, besides the old sedimentary and volcanic rocks, a small intrusion of granite around the State Fish Hatchery and excellent outcrops of serpentinite a few hundred yards east of Goodyears Bar.

Early miners worked several good placer deposits along this canyon but never found many worthwhile lode deposits. Downieville flashed briefly as a placer mining camp over a century ago and then limped along for years thereafter on a few small lode mines near town and a few others in the hills to the south.

Hydraulic miners left these boulders beside the Yuba River southwest of Goodyear's Bar.

DOWNIEVILLE — VINTON (on Hwy. 70)
(50 miles)

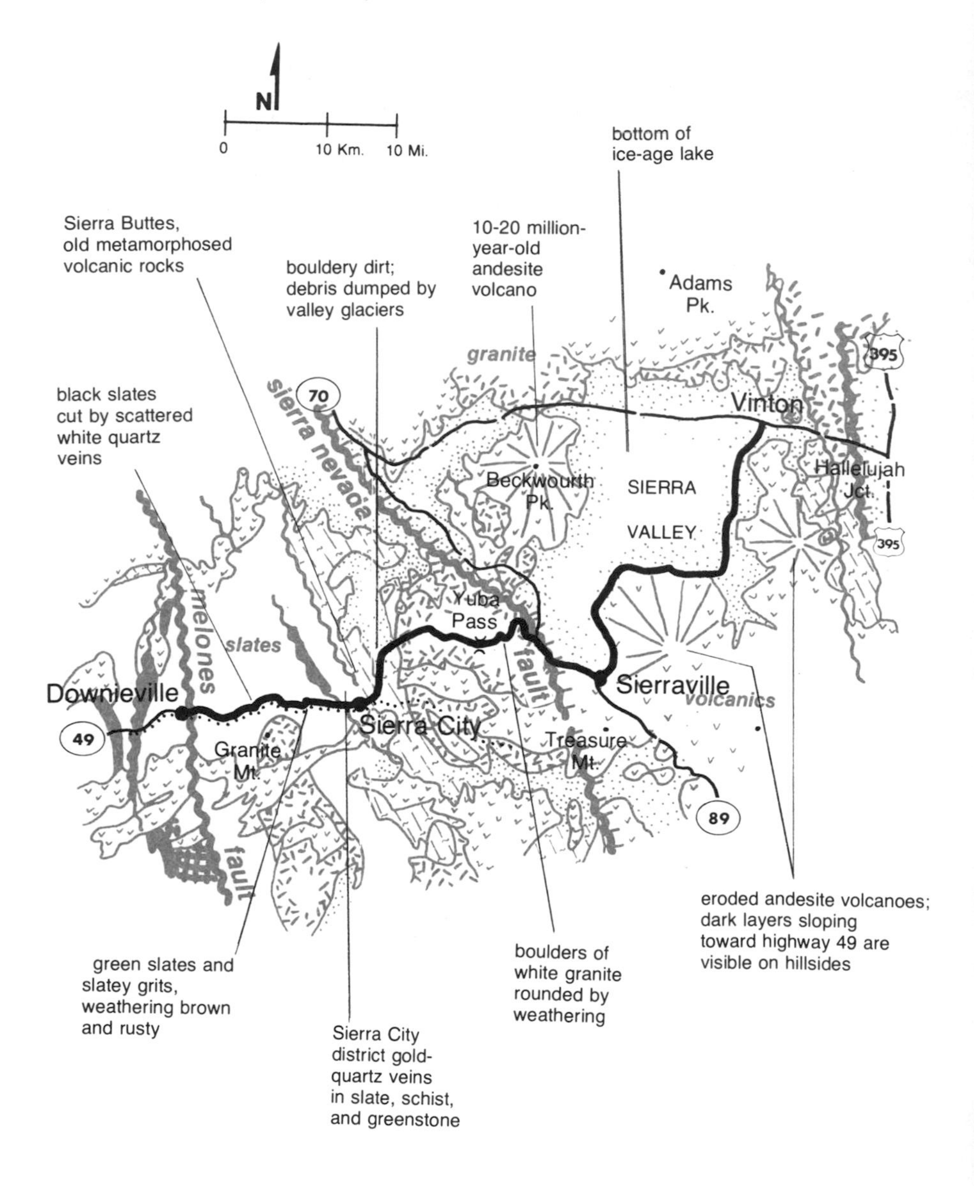

DOWNIEVILLE — VINTON

The route between Downieville and Vinton passes over the crest of the Sierra Nevada and into the dry Sierra Valley to the east. Between Downieville and Sierra City, a distance of about 13 miles, the outcrops are almost all sedimentary rocks, mostly dark slates but including a colorful variety of other kinds of rock. All have in common their origin as sea floor sediments swept together to make a coastal mountain range about 150 million years ago; all have been intensely deformed and considerably recrystallized by heating. White quartz veins streak a few of the roadcuts and water-sculptured outcrops along the river.

Immediately east of Sierra City the road passes on to a narrow belt of pink igneous rocks that contain large crystals of quartz and feldspar in a fine-grained matrix. About 8 miles northeast of Sierra City, at the community of Bassett, the highway enters the vast Sierra Nevada granite batholith, the geologic backbone of the range, leaving the gold country behind. Very little gold has ever been found within the granite.

Veins of white quartz in black slate near Downieville.

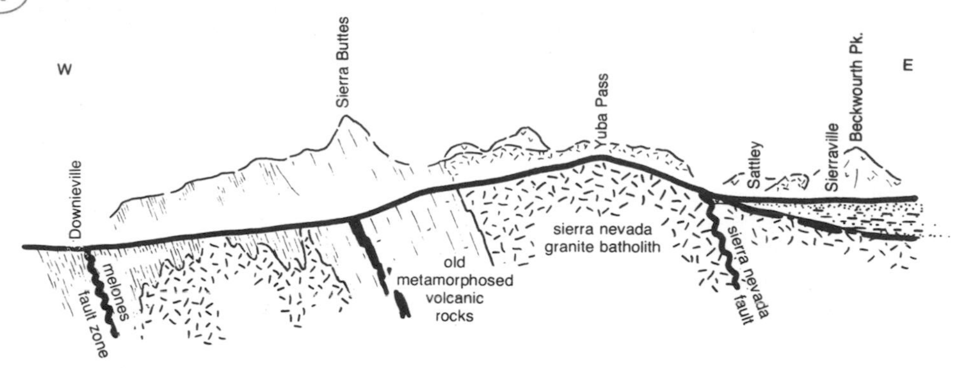

Section along the line of California 49 as it passes over the crest of the Sierra Nevada between Downieville and Vinton.

Sierra City is at the lowest limit of former glaciation in this valley, a boundary marked by large glacial moraines conspicuously visible in roadcuts near town. Moraines are hummocky heaps and ridges of debris dumped directly from the ice along the margin of a glacier; roadcuts through them reveal a disorderly mixture of boulders, gravel and sand all mixed together and left in a heap as though by a giant bulldozer. Ice-age glaciation has affected the scenery between Sierra City and Yuba Pass by scouring the valley into a broad and rather straight trough — in striking contrast to the narrow, winding canyon downstream.

Light grains of feldspar and quartz interlock with black crystals of hornblende to form granite. Shown natural size.

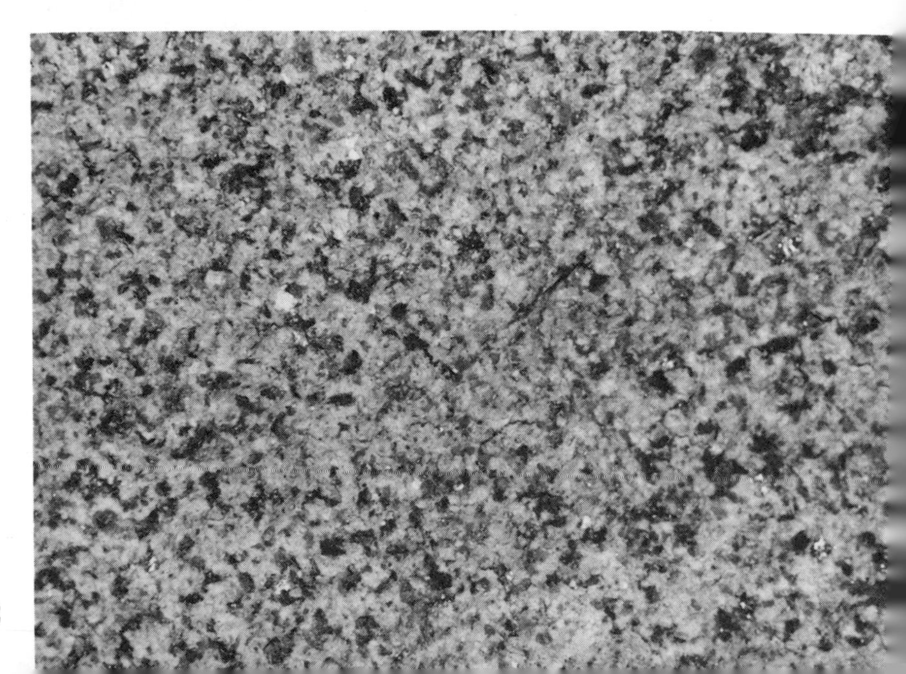

Weathered granite at Yuba Pass.

Yuba Pass is at the crest and eastern margin of the Sierra Nevada. East of the pass Highway 49 steeply negotiates the abrupt escarpment, a fault, that defines the eastern face of the range. Sattley and Sierraville are in the floor of the Sierra Valley, another large block of the continental crust that was dropped by the same faulting that raised the Sierra Nevada.

Between Sattley and Vinton, Highway 49 winds an angling course across the remarkably flat floor of the Sierra Valley, an old lake bed, skirting the northern margin of a patch of low wooded hills. These are eroded volcanic rocks that were erupted onto the floor of the Sierra Valley after it had been dropped by faulting. A few outcrops of granite that have been found poking through them in places show that the valley floor is underlain by the Sierra Nevada batholith — the same rock that forms the high mountain crest outlined on the skyline to the west.

OROVILLE — QUINCY (FEATHER RIVER CANYON)

(74 miles)

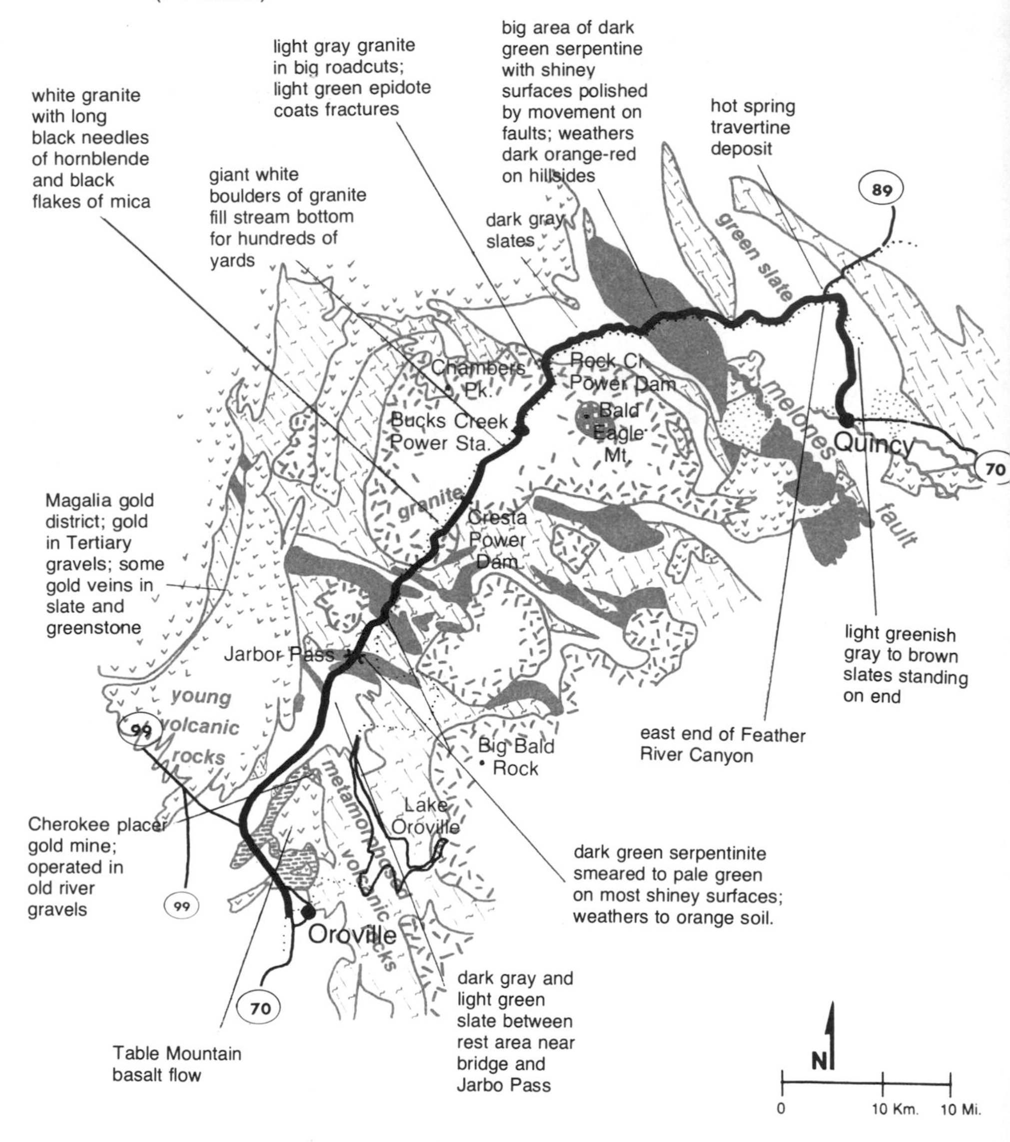

Table Mountain – its flat top is the surface of a lava flow that poured down a stream valley.

california 70 — feather river canyon

OROVILLE — QUINCY

Oroville is at the eastern margin of the Sacramento Valley on the thin edge of the young sedimentary deposits that fill the valley and lap onto the older rocks of the Sierra Nevada. Placer gold miners of the last century converted dozens of square miles of formerly lush floodplain downstream from Oroville into a bleak wasteland of tailings piles.

About 3 miles northwest of Oroville the new road passes through a gap in Table Mountain, one of California's best known geologic sights. It began as a valley eroded by a stream that once flowed out of the Sierra Nevada and onto the soft sedimentary rocks that fill the Great Valley. Then about 50 million years ago, volcanic eruptions poured a very large lava flow down the valley diverting the stream and paving its bed with a thick layer of basalt. Since then the softer rocks that made the old hills have eroded much more rapidly, leaving the hard basalt standing high above the surrounding countryside as the black rimrock capping Table Mountain.

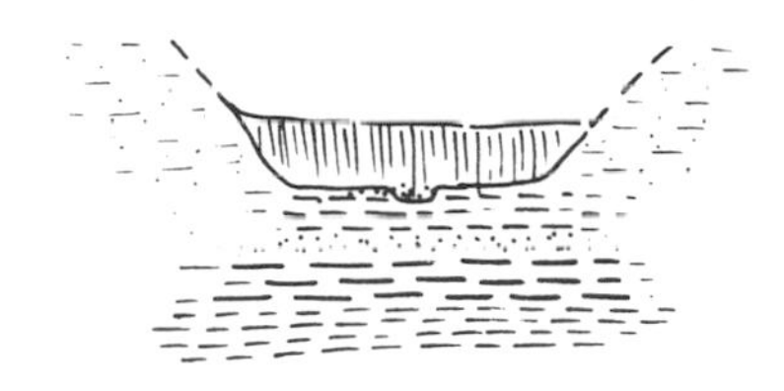

Table Mountain then and now. Erosion of softer rocks has left the lava-filled valley standing as a ridge.

North of Table Mountain the Feather River Road turns northeast, heading into the Sierra Nevada. Rocks west of Jarbo Pass began as soft sediments deposited beneath the waters of the Pacific Ocean about 200 million years ago and then metamorphosed as they were jammed into a marginal trench and heated nearly red hot about 150 million years ago. Now the old sediments are an assortment of rather dull-looking dark gray slates and buff-colored schists which contain enough fine-grained mica to glisten in the sun.

East of Jarbo Pass, along the slope between the pass and the canyon floor, the road passes several large bodies of greasy-looking serpentinite in various dark shades of green. Hardly any grass grows on the reddish-orange soils that develop on serpentinite and most shrubs do very poorly so the areas where it is exposed are easy to recognize even if no fresh rock is visible. Serpentinite, normally belonging in the earth's mantle beneath the ocean bottom, was incorporated into one of the big slices jammed into the continental

Sunlight glinting from shiny fracture surfaces in serpentinite. Just east of Rich Bar.

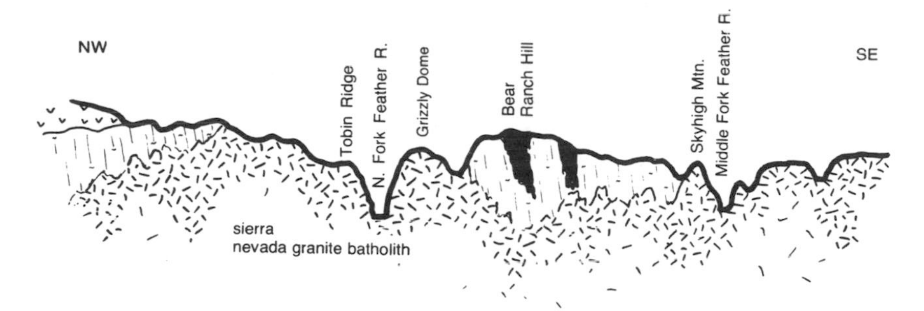

Section across Highway 70 in the Feather River Canyon, showing older rocks invaded by granites of the Sierra Nevada batholith.

margin. The rocks which contain these serpentinites are rather darkly nondescript. They consist mostly of volcanic debris that was dumped into the ocean and then swept into the marginal trench where it got hot enough to recrystallize into slabby-looking slates and schists.

For almost 20 miles between the Poe and Rock Creek power dams, the Feather River road passes through a large mass of granite. Large volcanoes almost certainly fed on this magma while it was still molten. Now that the volcanoes are eroded completely away, the magma that crystallized slowly beneath the surface without erupting is exposed as coarsely crystalline granite. Huge cliffs of pale gray and white granite make sheer canyon walls rising dramatically towards the edge of the plateau surface far above.

Slates beside the Feather River Road. Their slabby look results from closely-spaced fractures formed during metamorphism.

70

Boulders of stream sculptured granite in the Feather River near the Bucks Creek Power Station. They are remnants of a rockfall that dammed the river until it washed downstream.

The highway leaves the granite intrusion near the Rock Creek dam and passes through older sedimentary rocks rather similar to those west of Jarbo Pass along most of the route between there and Quincy. Like the others, these rocks suffered enough heating and general abuse as they were jammed into the marginal trench to convert them into dark slates hardly resembling the original muddy sediments. Some of them are called phyllites or schists if they contain enough mica to make the surfaces of freshly-cracked slabs glossy or glittery.

The broad belt of old sedimentary rocks is interrupted by a thick slab of old bedrock sea floor that originally lay beneath the sediments until everything was scrambled in the marginal trench. For a distance of several miles east of Rich Bar the road passes nearly continuous big cuts in dark green serpentinites and related black rocks. As usual, weathered hillslopes on the serpentinite are mantled with orange soils only thinly cloaked by a scrubby growth of shrubs. This particular slab of old sea floor extends for many miles along the length of the Sierra Nevada as a continuous belt following what geologists call the "Melones fault zone." It appears to have been one of the surfaces along which the descending slab of sea floor slipped beneath the continental crust. A number of other such surfaces have been recognized elsewhere in the Sierra Nevada and in the Coast Range. They become progressively younger westward because the marginal trench migrated westward through time.

QUINCY — HALLELUJAH JCT.
(58 miles)
N
0
10 Km.
10 Mi.

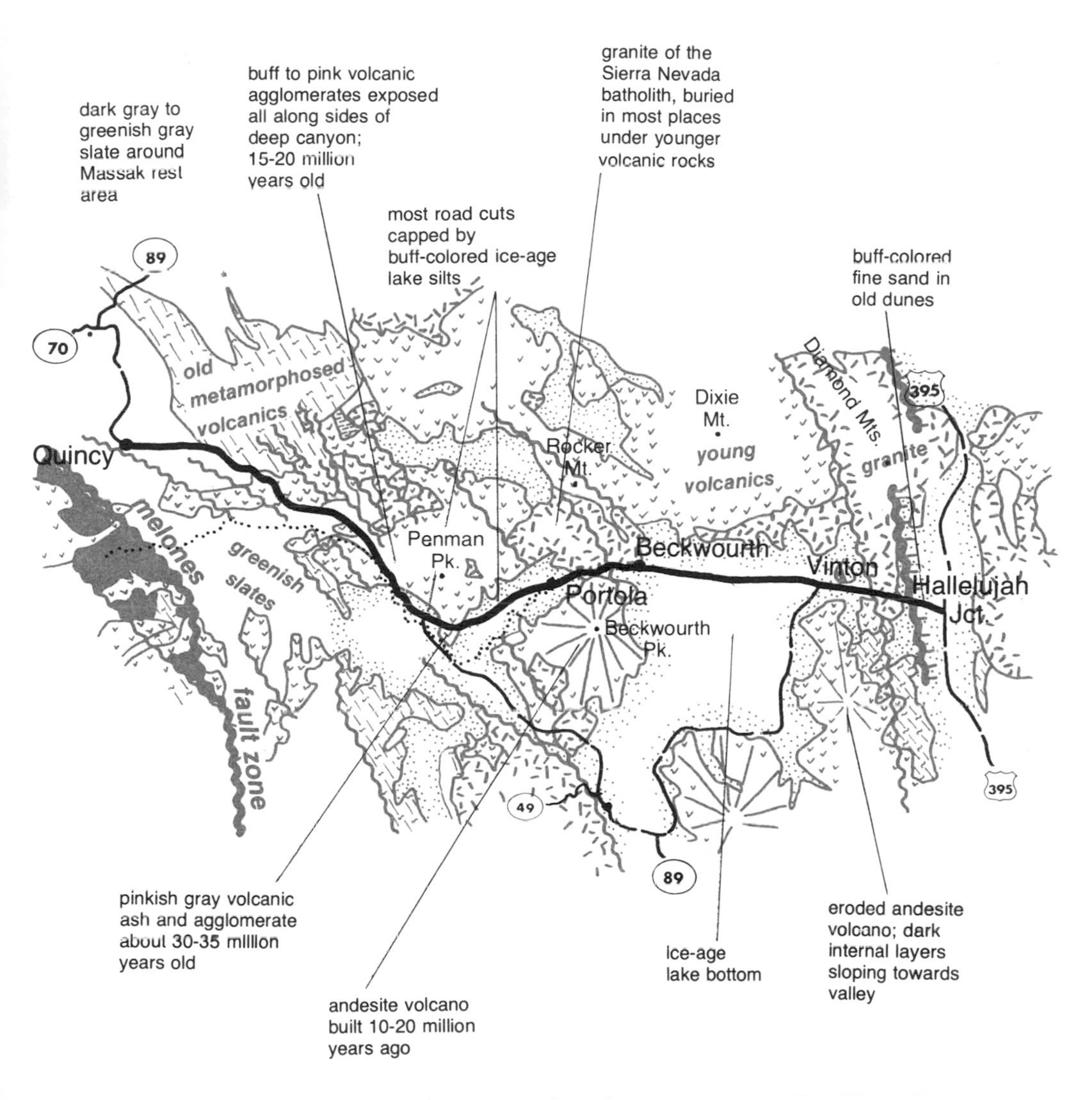
dark gray to greenish gray slate around Massak rest area
buff to pink volcanic agglomerates exposed all along sides of deep canyon; 15-20 million years old
most road cuts capped by buff-colored ice-age lake silts
granite of the Sierra Nevada batholith, buried in most places under younger volcanic rocks
buff-colored fine sand in old dunes
89
70
old metamorphosed volcanics
Quincy
melones
greenish slates
fault zone
Penman Pk.
Rocker Mt.
Dixie Mt.
young volcanics
Diamond Mts.
granite
395
Beckwourth
Portola
Beckwourth Pk.
Vinton
Hallelujah Jct.
49
89
395
pinkish gray volcanic ash and agglomerate about 30-35 million years old
andesite volcano built 10-20 million years ago
ice-age lake bottom
eroded andesite volcano; dark internal layers sloping towards valley

QUINCY — HALLELUJAH JUNCTION

Quincy is set in a broad valley that apparently formed by sinking of a small crustal block along a triangular set of faults. It is floored by deposits of loose sediment washed in from the surrounding mountains during the past several million years. Bedrock in those mountains began as sands and muds deposited in the Pacific Ocean 200 or 300 million years ago. Thorough heating and intense deformation that occurred as they were stuffed into a marginal trench about 150 or 200 million years ago converted them into the variety of dark gray and brown slates and schists that exist there today.

One of several big faults that define the eastern margin of the northern Sierra Nevada block passes beneath the eastern edge of the valley that contains Quincy. Highway 70 approximately follows this fault for most of the distance between Quincy and Blairsden. Its exact position is hard to know because the older bedrock along most of this route is buried beneath valley-fill sediments and young volcanic rocks.

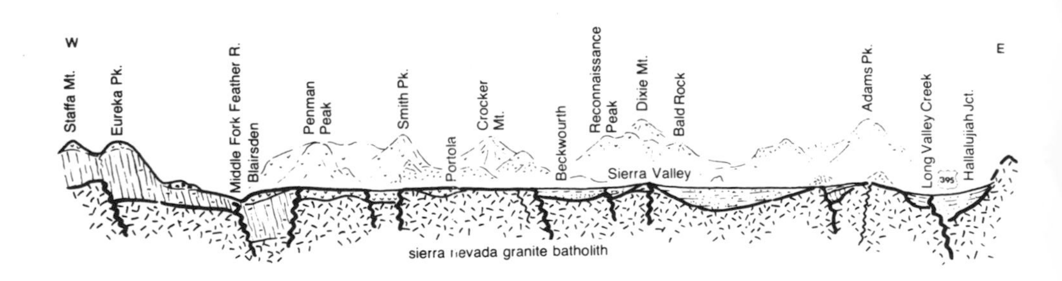

Section between the crest of the Sierra Nevada west of Blairsden and Hallelujah Junction.

Between Blairsden and Hallelujah Junction, California 70 crosses the floor of a broad basin created by general subsidence of the large crustal block immediately east of the Sierra Nevada. Rocks below the surface here are the same as those exposed on the high crest of the Sierra Nevada. But the bedrock basin is not clearly visible in the landscape because its outlines are confused by numerous groups of hills and small mountains. Some of these are granite — the large area that sank to form the basin is itself broken into smaller blocks along lesser faults and these rose and sank independently. Other hills on the basin floor are volcanic piles erected by eruptions that accompanied crustal movements of the last several million years.

There are large exposures of volcanic rocks beside the road where it passes through canyons north and south of Sloat and between Blairsden and Beckwourth. They consist mostly of ashfall and mudflow deposits in various pale shades of gray, pink and lavender. Many are agglomerates of volcanic ash mixed with angular chunks of volcanic rock.

Messy volcanic agglomerate of ash and angular chunks beside Highway 70 north of Blairsden.

Granites are exposed beside the road in the valley west of Beckwourth and they underlie the low, pine-covered hills north of the road from Beckwourth east. Plenty of granite is visible around Beckwourth Pass about 4 miles west of Hallelujah Junction, where the road passes over the southern tip of a large block of Sierran granite that stands prominently above the valley floor. Between Beckwourth Pass and the Junction, Highway 70 descends a long, sloping surface on gravelly sediments shed from the steep fault surface that formed the original east face of Diamond Mountain. From here north this fault forms the eastern front of the Sierra Nevada block. A few old sand dunes make buff-colored patches low on the slope.

Between Beckwourth and Beckwourth Pass, the highway crosses the extremely flat floor of the Sierra Valley, flat because it is underlain by lake-bed deposits. Evidently there was an undrained basin here trapped among the volcanoes and granite knobs. It filled with water, probably during the last ice age when rainfall was much heavier than it is now, and became a lake for a while until the outlet stream managed to erode its valley deep enough to restore drainage westward into the Feather River. The road follows this outlet valley between Beckwourth and Quincy.

It is interesting that the Feather River drains this area westward right through the high Sierra Nevada. Obviously the river is older than the outlines of the present landscape and managed to maintain its course through all the commotion of faulting and volcanism of the last several million years. This could not have happened unless the river was able to erode its channel downward more rapidly than the uplift of the Sierra Nevada block.

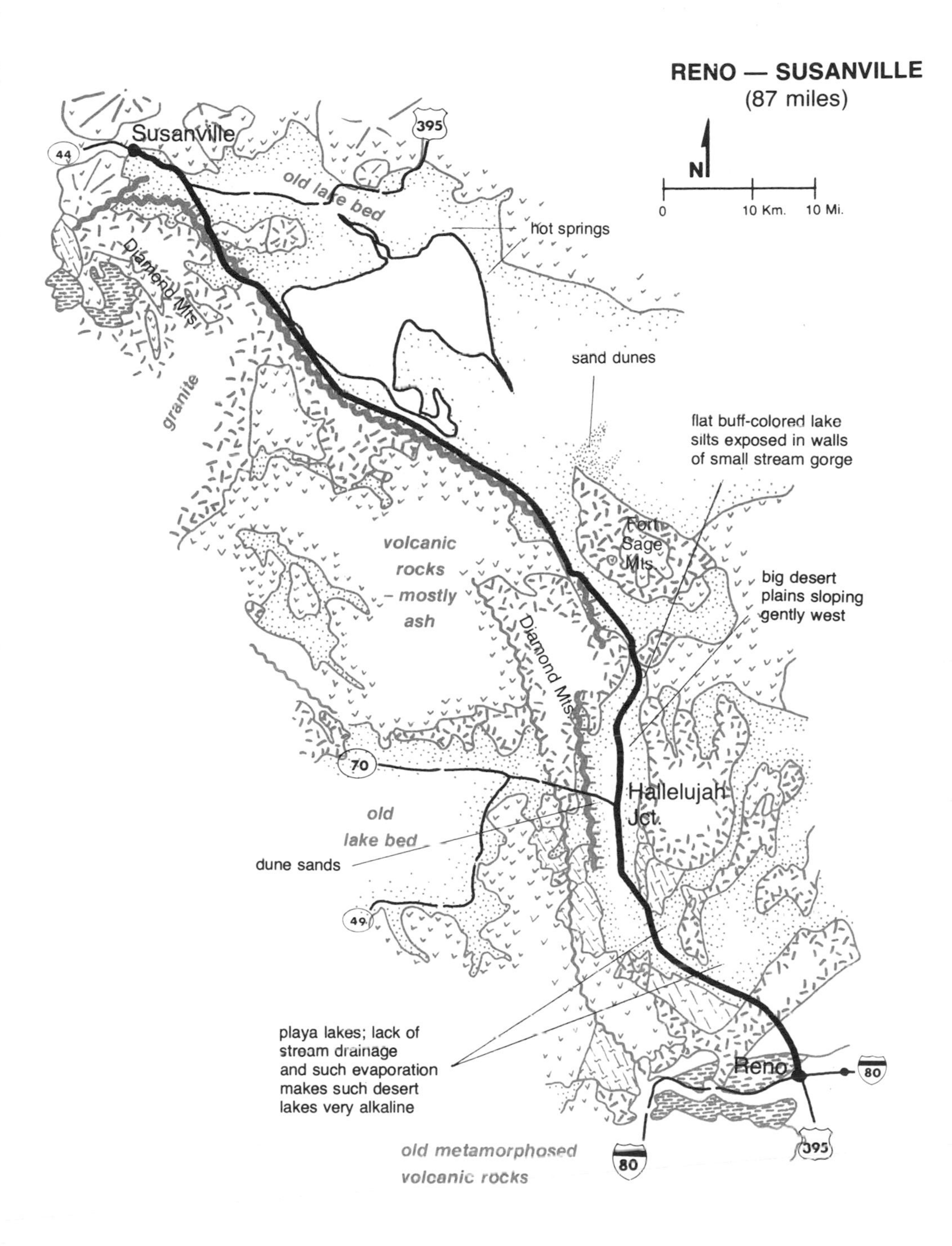
RENO — SUSANVILLE
(87 miles)
N
0
10 Km.
10 Mi.
Susanville
44
395
old lake bed
hot springs
Diamond Mts.
granite
sand dunes
flat buff-colored lake
silts exposed in walls
of small stream gorge
Fort
Sage
Mts.
volcanic
rocks
– mostly
ash
big desert
plains sloping
gently west
Diamond Mts.
70
Hallelujah
Jct.
old
lake bed
dune sands
49
playa lakes; lack of
stream drainage
and such evaporation
makes such desert
lakes very alkaline
Reno
80
395
old metamorphosed
volcanic rocks
80

u.s. 395

reno — susanville

The route between Reno and Susanville passes through valleys floored by recently deposited sediment separating mountains formed by a combination of block faulting and volcanic activity. The broad outlines of this landscape were established within the last few million years as the continental crust broke into large blocks which alternately rose and sank to become mountain ranges and basins. Widespread volcanic activity that accompanied the block faulting erected new volcanic mountains in the subsiding basins.

Bedrock in the fault-block mountains on either side of the road is almost all granite, probably related to that in the Sierra Nevada a few miles to the west. Bedrock in the young volcanic mountains consists largely of light-colored ash deposits and black basalt lava flows, along with a few other kinds of rock. Little bedrock is visible beside the highway which is routed through the valleys between mountains.

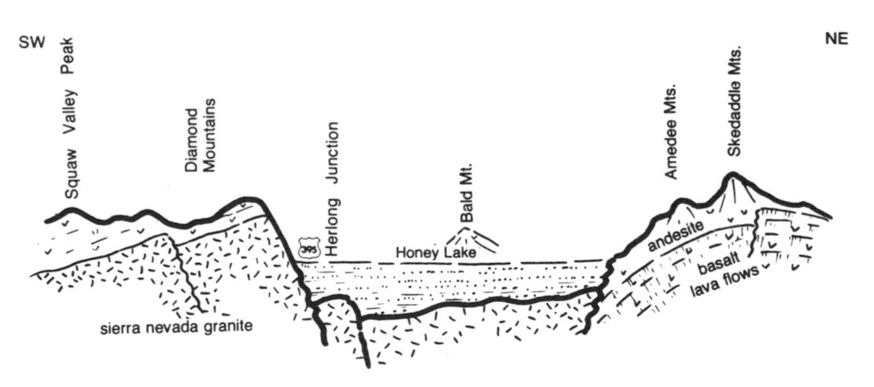

Section across the line of U.S. 395 just south of Honey Lake.

Desert landscapes have a distinctive look because the processes that shape them are different from those that dominate in areas with heavy vegetation. A very high proportion of the rain that falls in deserts runs off the surface because there are not enough plants to keep the soil open and absorbent. Heavy runoff that washes across desert slopes during the occasional heavy rains erodes numerous rills and gullies, giving the hills their look of having been intricately folded out of stiff cloth. Desert mountains rarely have the bulging humpy appearance so common in heavily forested regions where there is little surface runoff.

U.S. 395 heads south across an endless expanse of desert plain sloping gently down from the mountain front.

Where there are no leaves to absorb the force of their impacts, raindrops splash directly onto the soil, washing it away at an alarming rate. This explains why soil erosion is much more rapid in deserts than in humid regions where the ground is sheltered beneath a cloak of leafy plants. Desert mountains are usually picked nearly clean of soil leaving the rocky skeleton of the landscape exposed in harsh and jagged slopes. Geologists and artists generally regard this as a big improvement that contributes greatly to the interest and charm of the scene.

Rainsplashed soil washes off the rocky hillslopes and flushes down the gullies and canyons in the muddy flash floods that accompany heavy rains in the desert. These pour out of the mountains onto the valley floor depositing alluvial fans that coalesce as they grow larger to become broad aprons of sediment sloping gently downward from the mountain front to the valley floor. The same processes meanwhile erode the surface back into the bedrock of the mountains. Viewed from points in the valley, as from U.S.

395, these depositional and erosional desert plains appear as smooth surfaces rising gently toward the mountain front which rears abruptly above them. Geologists have been fascinated by their smoothly mathematical form for over a century and dispute endlessly over the details of their origin.

Deserts don't receive enough water to keep their streams flowing between rains or to enable them to flow very far when it does rain. Evaporation and loss of water into the ground cause most desert streams to shrivel and disappear before they have flowed beyond the nearest large valley. Long Valley Creek, which U.S. 395 follows along half the route between Reno and Susanville, empties into Honey Lake which is alkaline and salty because it has no outlet. Evidently evaporation from the lake balances inflow of water from Long Valley Creek and other tributary streams.

Most desert streams are muddy and obviously the mud can go no farther than the water. But mud doesn't evaporate so desert valleys fill with a continually deepening accumulation of sediment, slowly submerging the mountains in their own debris.

The ice ages, of which there were at least four during the last few million years, were periods of very heavy rainfall. Most of the desert lakes grew much larger then and undoubtedly became much fresher. Old shorelines, relics of lake levels during the last ice age, are visible along the margins of almost every desert valley as perfectly horizontal lines faintly grooved into the base of the hills.

REDDING — DUNSMUIR
(55 miles)

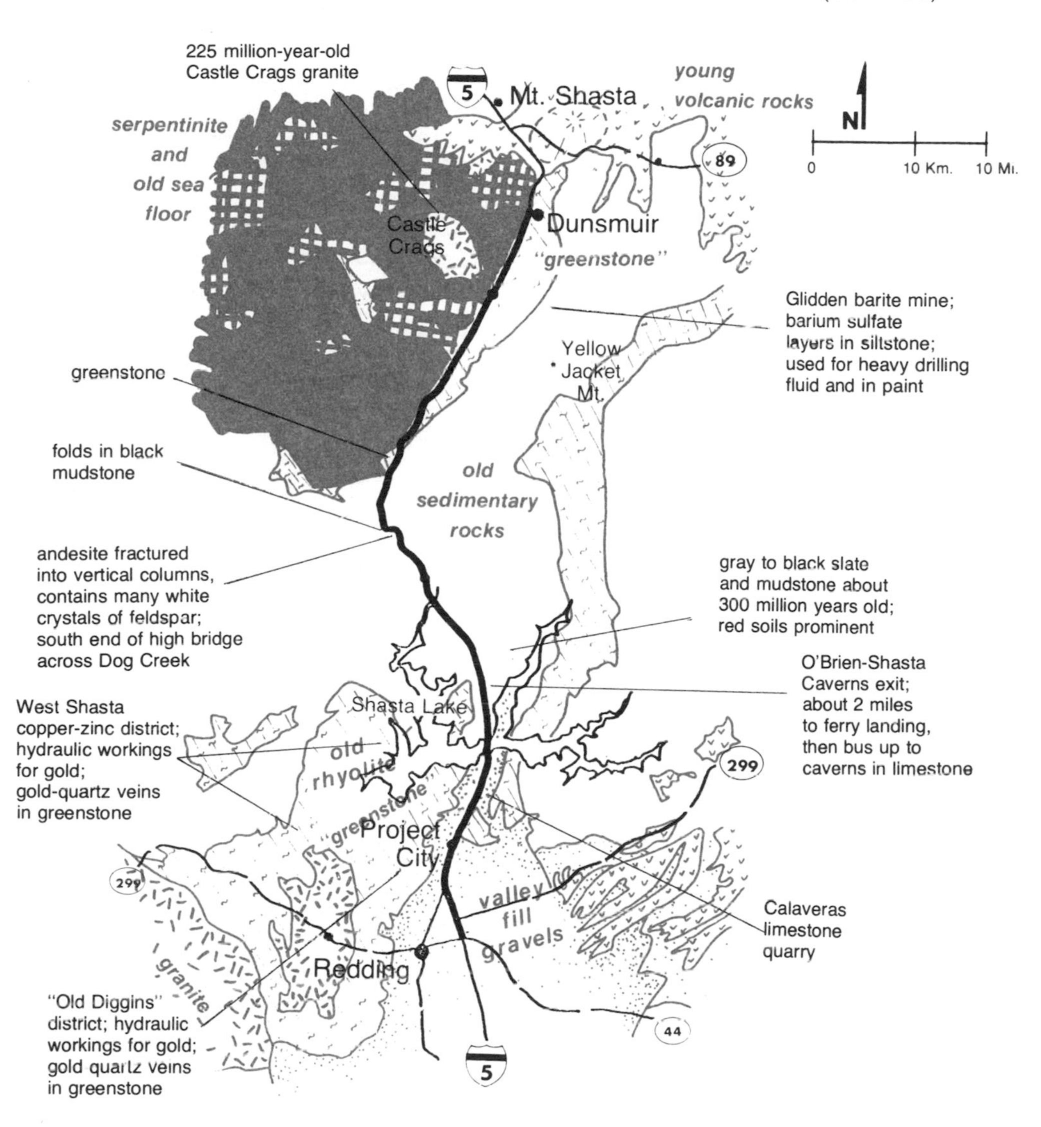

interstate 5
redding — dunsmuir

Redding is in the northernmost end of the Sacramento Valley where it butts against the abrupt southern wall of the Klamath Mountains — the broken edge that matches the northern edge of the Sierra Nevada. The entire route between Redding and Dunsmuir passes through these mountains across rocks essentially similar to the older bedrock in the northern Sierra Nevada.

Rocks of the Klamaths are poorly exposed, poorly understood, and very complicated. These rugged mountains are extremely frustrating to geologists because most of their rocks are nearly hidden beneath deep soils and lush greenery. Occasional exposures offer tantalizing glimpses of fascinating rocks that aren't exposed anywhere else within miles — if there is such a thing as a geologic strip tease, this is it. There is every likelihood that valuable mineral deposits are hidden in the Klamaths so the future will probably see determined efforts to unravel the geologic secrets of these hills.

Interstate 5 follows the valley of the Sacramento River along the entire route between Redding and Dunsmuir. As often happens, the river has followed the easiest course along the trend of the bedrock so essentially the same kinds of rocks are exposed for long distances along the route of the highway. Between Redding and Castella the valley is cut into old sedimentary rocks deposited on the floor of the Pacific Ocean as much as 300 or more million years

Andesite lava flow broken by shrinkage fractures into vertical columns that stand like a thicket of stumps beside I-5 south of Castle Crags.

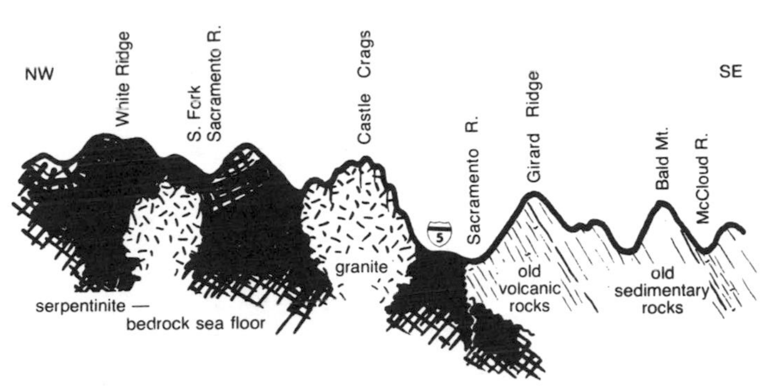

Section across I-5 showing granite intrusions punched through rocks that once floored the Pacific Ocean.

ago and then swept into the marginal trench sometime around 150 to 200 million years ago. These are some of the oldest rocks in northern California. Between Castella and Dunsmuir the bedrock is mostly black igneous rock and serpentinite that was once the bedrock floor of the ocean before it became mixed with the sediments in the scrambling confusion of the marginal trench.

Red soils are very conspicuous for long distances along the route north of Redding. They testify to a long period of weathering under warm and rather wet climatic conditions, the kind of situation in which such soils always form. Millions of rainy years washed everything soluble out of these soils leaving a sterile residue of simple clays stained red by iron oxide. This kind of soil is usually good for growing trees but not for crops unless it is heavily dosed with fertilizer.

Castle Crags, just west of the highway a few miles south of Dunsmuir, is an eroded granite intrusion protruding jaggedly as though it were the splintered stump of an enormous tree. Towering pillars and monster boulders of granite with trees growing among them suggest to many people the picturesque ruins of a fantastic fortress. Castle Crags is actually no different from dozens of other granite intrusions in the Klamaths and Sierra Nevada except that it has lost most of its soil cover to erosion. Granites weather to soil most rapidly along the widely spaced fractures that naturally divide the rock into blocks. Solid rock between the fractures weathers more slowly and remains as rounded masses of fresh rock embedded in the soil. If the soil should happen to be washed away, probably as the result of damage to its sheltering cover of plants, the unweathered remnants of fresh rock are left behind as exposed boulders and pillars of granite.

ARCATA — JUNCTION CITY

(89 miles)

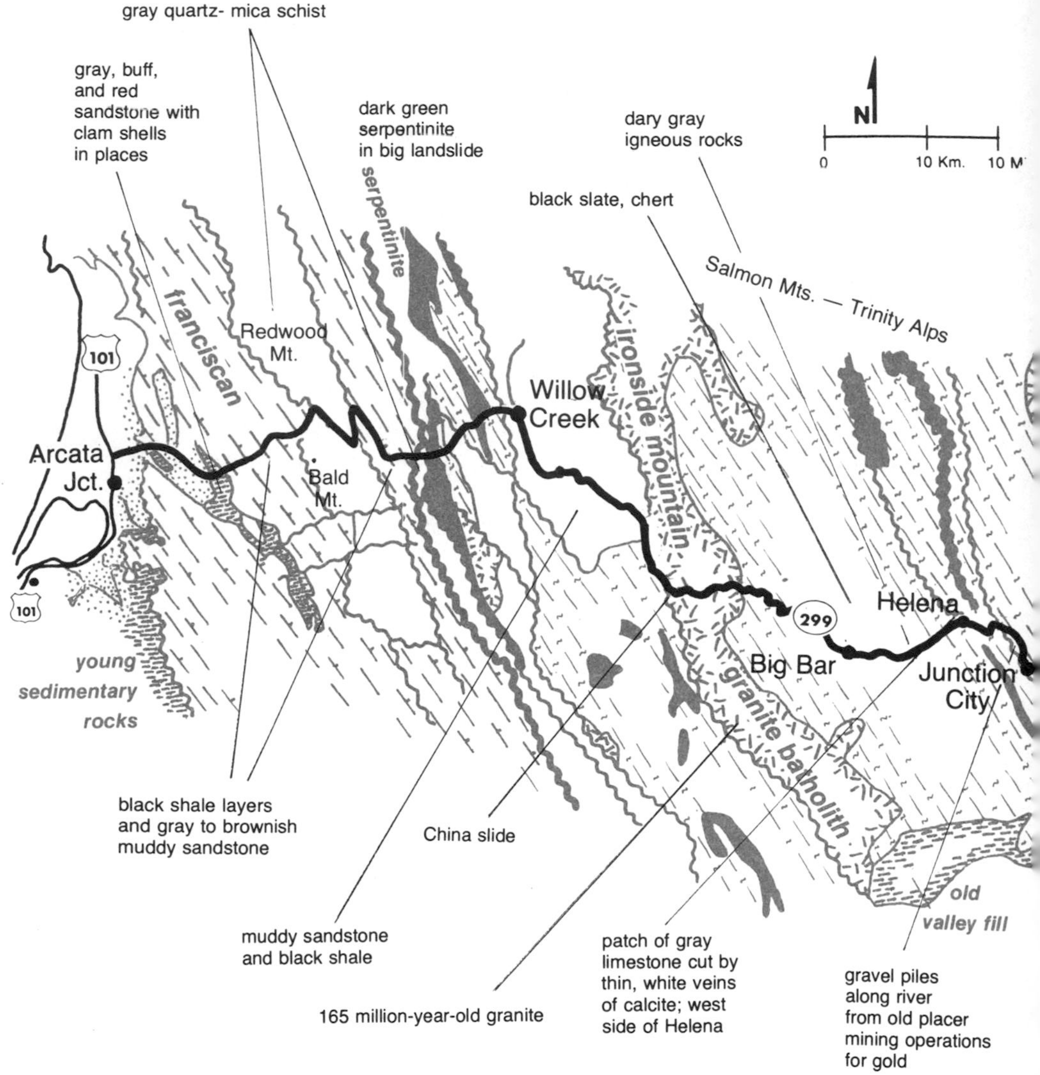

gray quartz- mica schist

gray, buff, and red sandstone with clam shells in places

dark green serpentinite in big landslide

dary gray igneous rocks

N

0 10 Km. 10 M

black slate, chert

Salmon Mts. — Trinity Alps

serpentinite

franciscan

Redwood Mt.

101

Willow Creek

ironside mountain

Arcata

Jct.

Bald Mt.

101

299

Helena

young sedimentary rocks

Big Bar

Junction City

granite batholith

black shale layers and gray to brownish muddy sandstone

China slide

old valley fill

muddy sandstone and black shale

patch of gray limestone cut by thin, white veins of calcite; west side of Helena

gravel piles along river from old placer mining operations for gold

165 million-year-old granite

california 299 — the trinity highway

ARCATA — JUNCTION CITY

Between Arcata and Redding, U.S. 299 winds through the northern Coast Range and the Klamath Mountains, through a thinly peopled part of California that has received remarkably little attention from geologists. It is rugged, nearly inaccessible country densely covered with forest and brush and sadly lacking good natural outcrops. Even though resources of platinum, chromium, gold, manganese and other valuable minerals are known to exist in these wild mountains, the problems of development and transportation have been forbidding enough to discourage much interest in geological exploration. Men of the U.S. Geological Survey have been making strenuous efforts in recent years to fill in some of the worst gaps in our knowledge of the complex but intriguing geology hidden in these mountains.

All of the rocks in the northern Coast Range and most of those in the Klamaths began as sediments deposited on the floor of the Pacific Ocean and later scrapped into a marginal trench and onto the edge of the continent. The Klamaths are geologically distinct from the Coast Range in that they contain rocks that are somewhat older and have been extensively intruded by large bodies of granite — they are the northern extension of the Sierra Nevada. Nevertheless, the rocks in the Klamaths and Coast Range are very similar and the exact boundary between them has been a point of minor dispute among geologists. There is no way to see the difference by looking at the landscape.

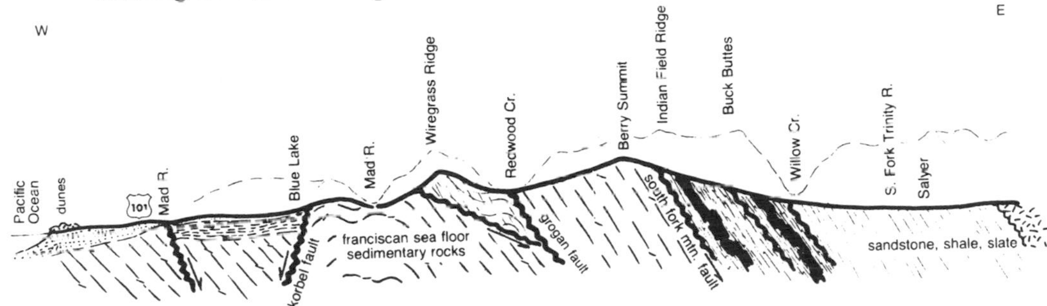

Section along U.S. 299 between Arcata and Willow Creek. Younger rocks in the west were jammed beneath older ones to the east as they were stuffed into a marginal trench.

As good a place as any to draw the line separating the Coast Range and the Klamaths is the western base of Indian Field Ridge, about 5 miles west of Willow Creek. Between there and Arcata, U.S. 299 winds through hills underlain by Franciscan rocks, nondescript mudstones and muddy sandstones like those seen everywhere in the Coast Range. Some of them were once hot enough to cook a bit, especially those in Redwood Mountain, and show the effects of slight metamorphism in their slabby fracture pattern and glistening broken surfaces.

In the westernmost fringe of the Klamaths, between the base of Indian Field Ridge and the community of Willow Creek, U.S. 299 passes through several small belts of heavy, greenish-black rock. Formidable outcrops of this thoroughly sheared rock were once part of the bedrock floor of the Pacific Ocean — actually the top of the earth's mantle — and somehow got scrambled into the Klamaths while the sea floor was sliding into a marginal trench here 150 or 200 million years ago. Muddy sandstones in the same area are the sediments dumped on the same sea floor.

There happens to be a great lot of this old black sea floor in the Klamaths, much more than in the Coast Range or the Sierra

Trinity River gorge exposes a deep slice through the old sea floor rocks in the Klamaths.

Nevada, and it gives these mountains much of their distinctive geologic character. Such rocks always contain chromium and platinum and both have been mined casually in the Klamaths for many years, the chromite in small pockets of high grade hardrock ore and the platinum along with gold in stream placer deposits. Larger deposits of both almost certainly await discovery.

An old hydraulic nozzle used at the Lagrange Mine.

U.S. 299 follows the canyon of the Trinity River between Willow Creek and Junction City. Mudstones and muddy sandstones exposed in this area were deposited earlier and scraped off the sea floor sooner than those farther west but look much the same. Between the communities of Burnt Ranch and Del Loma, the road passes through the Ironside Mountain batholith, a large body of granite that shouldered its way upward into the sedimentary rocks as a mass of molten magma about 165 million years ago. Granites in the Klamath Mountains are similar to those in the Sierra Nevada but not nearly so numerous. In both ranges they appear to have been responsible for development of lode deposits of gold. The red hot intrusion of molten granite forced circulation of hot water and steam through the nearby rocks dissolving gold and quartz at depths of more than two miles and depositing them again as bold-bearing quartz veins nearer the surface.

JUNCTION CITY — REDDING
(58 miles)

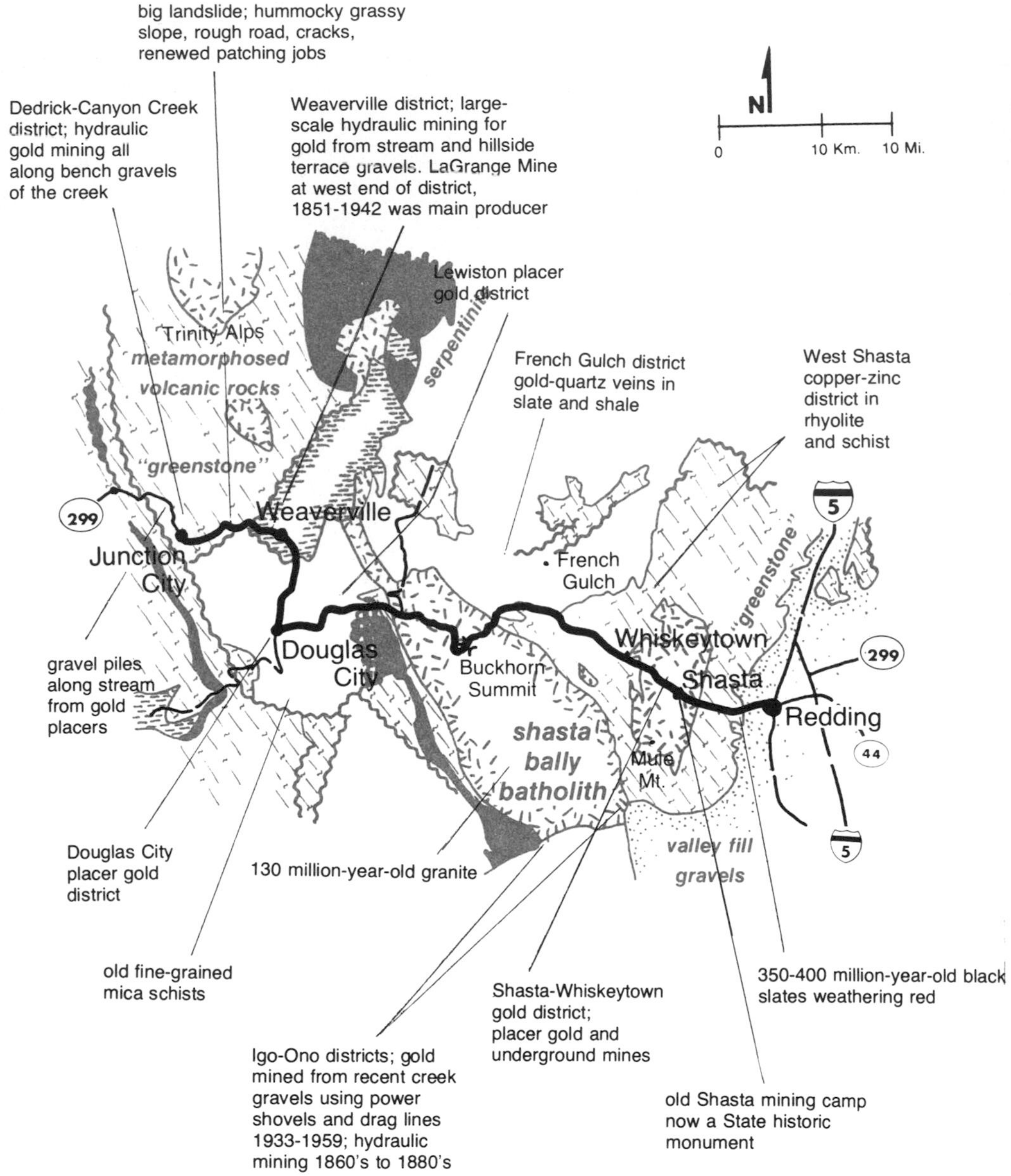

big landslide; hummocky grassy slope, rough road, cracks, renewed patching jobs

Dedrick-Canyon Creek district; hydraulic gold mining all along bench gravels of the creek

Weaverville district; large-scale hydraulic mining for gold from stream and hillside terrace gravels. LaGrange Mine at west end of district, 1851-1942 was main producer

N

0 10 Km. 10 Mi.

Lewiston placer gold district

Trinity Alps

metamorphosed volcanic rocks

serpentinite

"greenstone"

French Gulch district gold-quartz veins in slate and shale

West Shasta copper-zinc district in rhyolite and schist

299

Weaverville

Junction City

5

French Gulch

"greenstone"

Whiskeytown

Douglas City

Buckhorn Summit

Shasta

299

Redding

gravel piles along stream from gold placers

shasta bally batholith

Mule Mt.

44

valley fill gravels

5

Douglas City placer gold district

130 million-year-old granite

old fine-grained mica schists

Shasta-Whiskeytown gold district; placer gold and underground mines

350-400 million-year-old black slates weathering red

Igo-Ono districts; gold mined from recent creek gravels using power shovels and drag lines 1933-1959; hydraulic mining 1860's to 1880's

old Shasta mining camp now a State historic monument

Slabby outcrop of schist. A closer look would show that the rock has a streaky "grain" because its mineral crystals are as regularly aligned as bricks in a wall.

JUNCTION CITY — REDDING

East of Junction City the road rises out of the canyon to cross hills eroded into sedimentary rocks that were heated enough to get cooked into metamorphic schists that hardly resemble the original muddy sediments. They contain big spangly crystals of black or white mica that sparkle in the sun and needle-shaped crystals of glossy black hornblende. Unlikely as it seems, these crystalline rocks have about the same chemical composition as the muddy sediments common elsewhere in the Klamaths and the Coast Range.

Weaverville is in a depression deeply floored by loose sands and gravels eroded from the surrounding hills and washed in during the last several million years. Between Weaverville and Redding the road crosses another large area of schist and two granite batholiths. Granites look very much alike regardless of where they are or how old they may be — massive light gray or pink rocks widely seamed by fractures and veins. Beyond the roadcuts, their outcrop areas are usually weathered into large, rounded boulders that suggest giant sofa pillows littering the countryside.

A gallery of dikes

Molten granite magma squirted into fractures in older metamorphic rocks.

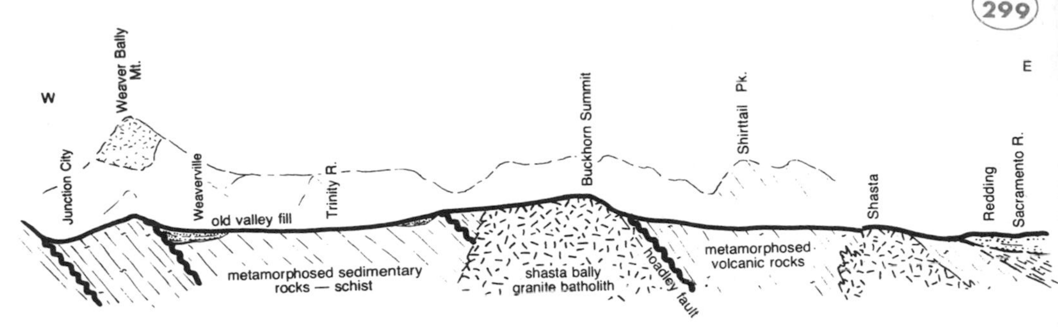

Section along U.S. 299 between Junction City and Redding.

Redding is in the Sacramento Valley just south of the torn southern boundary of the Klamath Mountains. Rocks in the northernmost Sierra Nevada about 60 miles east of Redding match those in the mountains directly north of town.

CROSS-REFERENCES

More about gold mining in Chapter III, U.S. 50, California 49 and California 70.

More about the origin and distribution of gold in the introduction to Chapter III, Sierra Nevada and the Klamaths.

More about the crest of the Sierras in Chapter III, Interstate 80, Sacramento — Reno.

More about gold dredging in Chapter III, U.S. 50: Sacramento — Carson City and California 70; Oroville — Quincy.

More about the origin and distribution of gold in the intruduction to Chapter III, Sierra Nevada and the Klamaths.

More about gold dredging in Chapter III, U.S. 50: Sacramento — Carson City

More about gold mining in Chapter III, California 49: Placerville — Vinton.

More about old sea floor and serpentinite in Chapter I and in the introduction to Chapter II, The Coast Ranges.

More about old sea floor in Chapter I, The Great Collision and in the introduction to Chapter II, The Coast Ranges.

IV

the great valley

— a trough of mud

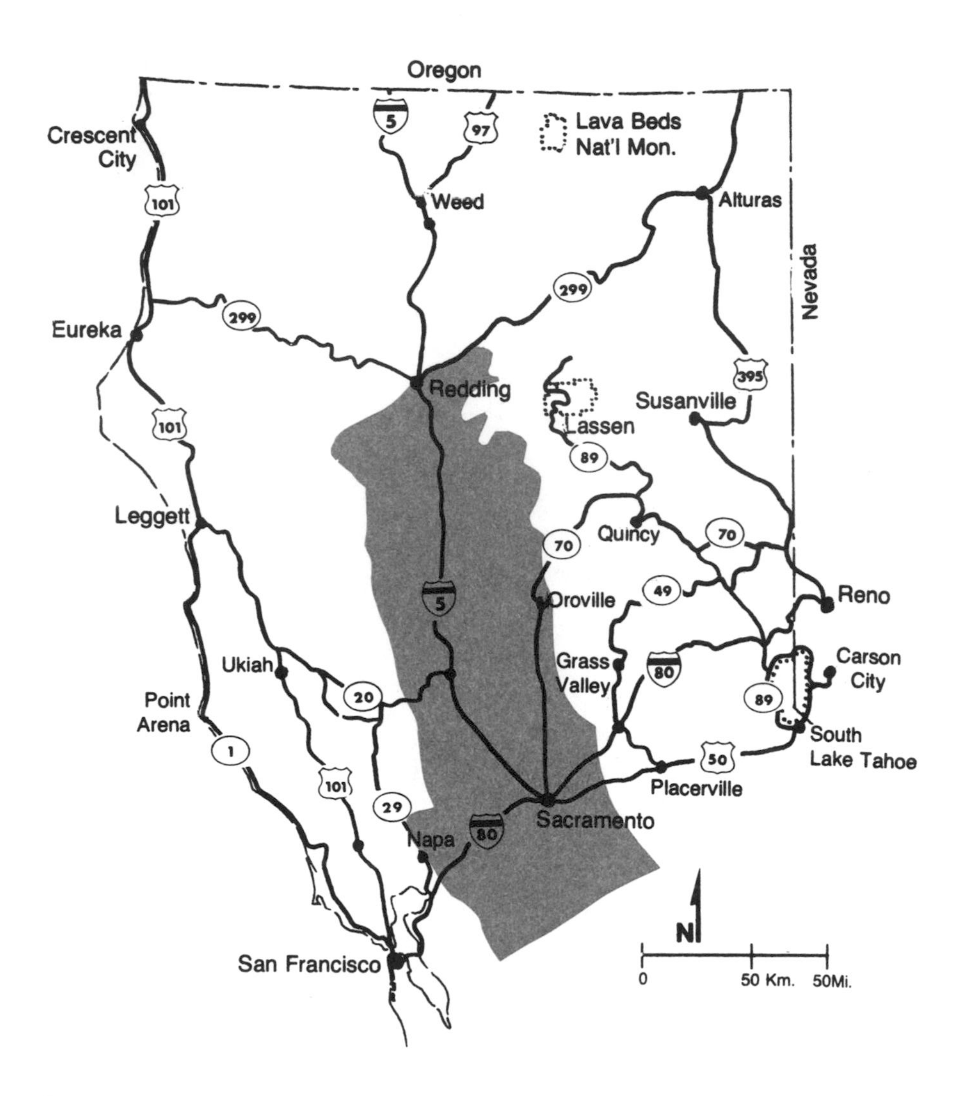
Oregon
Lava Beds
Nat'l Mon.
Crescent
City
Weed
Alturas
Nevada
Eureka
Redding
Susanville
Lassen
Leggett
Quincy
Oroville
Reno
Ukiah
Grass
Valley
Carson
City
Point
Arena
South
Lake Tahoe
Placerville
Sacramento
Napa
San Francisco
N
0
50 Km.
50Mi.
5
97
101
299
395
89
70
49
80
20
1
29
50

— a trough of mud

Enclosed between the Sierra Nevada on the east and the Coast Ranges on the west, the Great Valley is a long, narrow trough once filled with seawater and now with muddy sediments accumulated over millions of years. Its northern end abuts the truncated southern edge of the Klamath Mountains.

Exactly what the Great Valley is, precisely how and when it originated, is a mystery made nearly impenetrable by those thousands of feet of mud in its floor. Rocks that might provide evidence leading to direct answers for those questions are deeply buried. Since the geologic secrets of its origin are so well hidden, geologists are reduced to speculation and differences of opinion based on fragmental evidence.

It seems likely that the Great Valley first became an isolated arm of the sea sometime near the middle part of Cretaceous time, perhaps about 140 million years ago. That is about the time that the Klamath Mountains separated from the northern Sierra Nevada and about the age of the oldest valley-filling sediments clearly deposited within the trough.

Exactly why the Great Valley exists is not completely understood. Troughs on the landward sides of mountain ranges formed by stuffing sea floor sediments under the edge of the continent also exist in the Phillipine Islands, Japan, the Aleutians and various other places. The trough seems to be simply a remnant of the old sea floor isolated by the new mountains and filled with sediments.

However it may have originated, the Great Valley did fill with sediment. The first muds were deposited in seawater, the older formations laid down during Cretaceous time contain numeous fossils of animals that lived in the ocean. By the early part of the Tertiary Period, about 50 million years ago, the water in the Sacramento Valley seems to have become quite shallow, even though deep water still existed farther south, and large parts of the valley were filled to sea level and receiving muds deposited above sea level. Since then, various parts of the Sacramento Valley have been either just below or just above sea level and accumulating sediments deposited either in shallow sea water or on dry land. Of course this is the situation today, most of the valley floor is slightly above sea level except for San Francisco Bay which is slightly below.

About 1½ million years ago, a time when the Sacramento Valley had already filled to sea level and become more or less dry land, the peace of the late Tertiary Period was briefly punctuated by a noisy series of volcanic eruptions at Sutter Buttes. They built a large volcano on the flat floor of the valley and its eroded remnants remain today as an isolated patch of hills ornamenting an otherwise flat valley floor.

Natural gas and some petroleum are produced in considerable quantity from the sedimentary rocks filling the Great Valley. These valuable commodities are likely to occur wherever thick sequences of sedimentary rocks have been deposited in sea water and the Great Valley is just the sort of place where they would be expected.

SAN FRANCISCO BAY — SACRAMENTO
(69 miles)

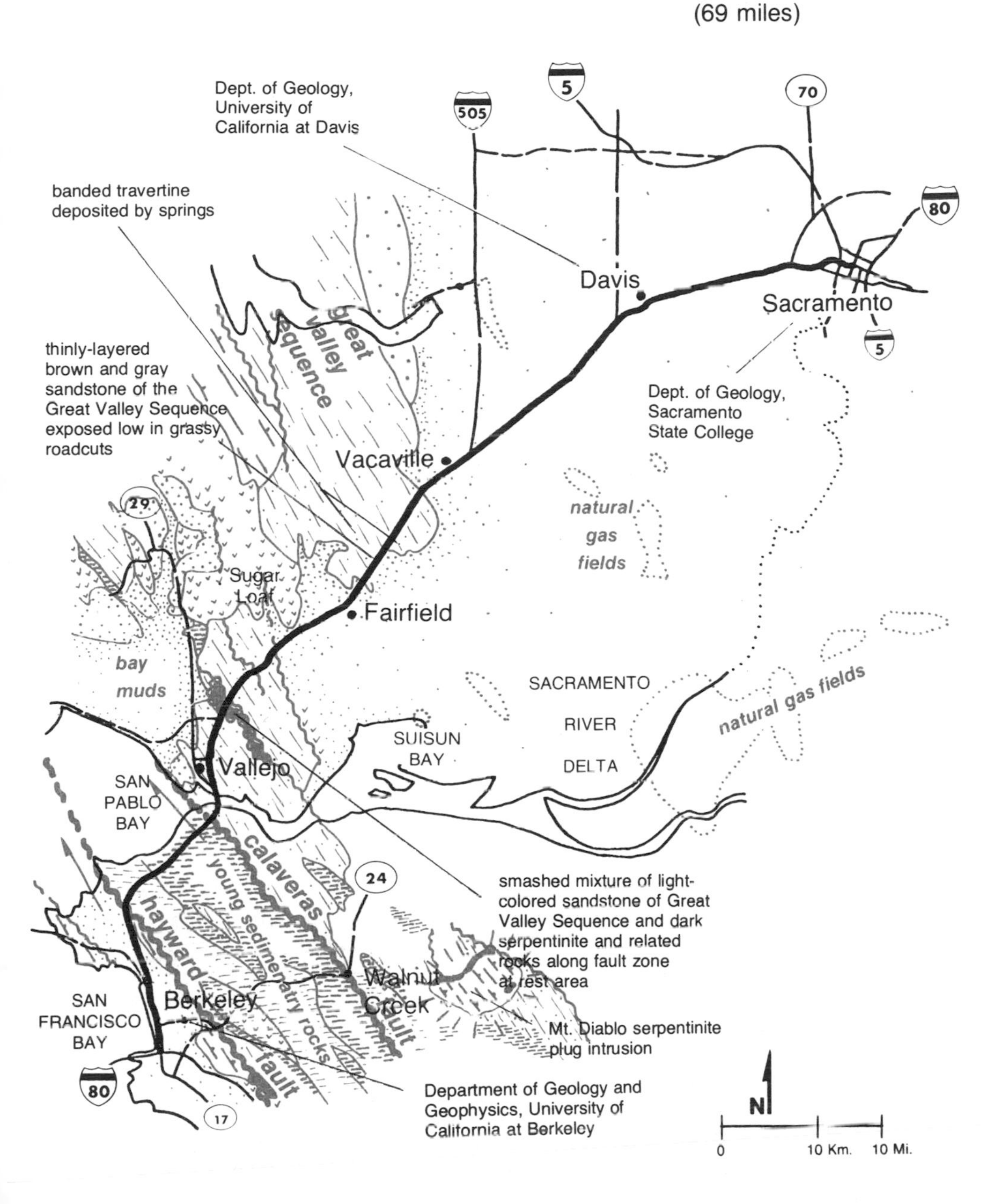

interstate 80
bay area — sacramento

Interstate 80 crosses the Hayward fault about where it leaves the eastern limits of Richmond to climb into the low hills that face the southeast side of San Pablo Bay. As it happens, the Hayward fault is one of those geologic features that clearly show on ordinary road maps — it defines the straight inland edge of the east bay urban complex. Low and flat alluvial ground between the fault and the bay was more easily developed and is now far more densely populated than the steep hills eroded into higher ground east of the fault. The south approach to the big bridge over Carquinez Straight is built across the Calaveras fault, another active member of the San Andreas system of faults.

Rocks that form the hills between Richmond and Vallejo are sedimentary formations containing fossils of animals that lived in the sea about 20 million years ago. Somewhat similar, but older, formations are exposed along several miles of the road in the hills between Vallejo and the junction with Highway 21. Evidently this part of the Coast Range has been submerged at various times since the mountains were formed.

These younger formations contain layers of all sorts of sediment ranging from rather hard sandstone and volcanic ash to soft muds. Such rocks are never very solid even under the best of circumstances and here the worst possible situation exists because crustal movements have folded the rocks so their layers are steeply tilted. Imagine putting a stack of heavily buttered pancakes on a plate, the pancakes representing the harder layers and the butter the layers of soft mud, and then suppose that you tilted the plate.

The natural situation plainly visible from Interstate 80 is about what anyone would expect of the imaginary experiment — everything is sliding on the soft layers. Nearly every hill underlain by

the younger sedimentary formations is scarred and wrinkled by landslides, old ones that slumped centuries ago and modern ones moving today. Nothing looks very stable for reasons that are sufficiently obvious wherever a roadcut exposes the steeply tilted thin layers of solid and weak sedimentary rock. It is very difficult to prevent such landslides because they are caused by naturally weak bedrock. These hills are destined to become a major headache if future population pressures force dense urban development.

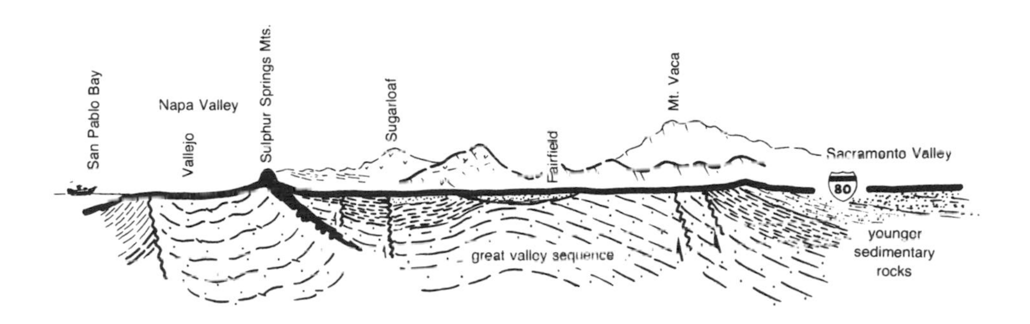

Section between Vallejo and Vacaville, looking north from I-80.

Between Vallejo and Vacaville, Interstate 80 passes the south ends of several long finger ridges, separated by wide valleys, that extend to the northern horizon. These are slices of the Coast Range moved northward by displacements along a system of faults that trend parallel to the San Andreas and are probably related to it. Except for some knobs of volcanic rock both north and south of the road near Rockville, bedrock in all these ridges consists mostly of dark, muddy sandstones belonging to the Great Valley Sequence. Ordinarily, these form a single continuous ridge along the eastern flank of the Coast Range; evidently faulting in this area has made one ridge into several.

The low ridge at Vacaville is the easternmost outpost of the Coast Range. Between there and Sacramento the highway crosses stream deposits laid down on the flat floor of the Great Valley within the very recent geologic past. On clear afternoons the Sierra Nevada is visible in the distance, looking like cardboard stage scenery pasted onto the eastern horizon. The Coast Range looks more convincing with its succession of north-south ridges fading into the blue distance of the hazy coastal air to the west.

interstate 5
sacramento — redding
— following the long trough

All the way between Sacramento and Redding, nearly 160 miles, Interstate 5 follows the floor of the Sacramento Valley only a few feet above sea level and nearly as flat as the sea itself. On a clear day, the Coast Range forms an irregular low horizon in the west and the distant peaks of the Sierra Nevada are silhouetted against the eastern skyline. The entire width of the Sacramento Valley is visible from the road and the geologic story is essentially the same all the way.

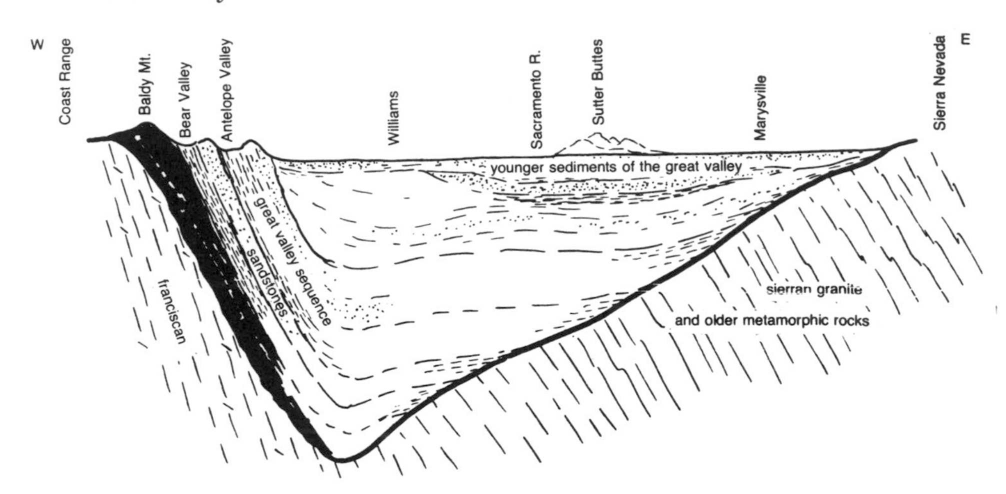

Section across the Great Valley. It really is a trough.

Loose sediments, mud washed into the valley from the surrounding mountains during the last few hundred thousand years, are all that is exposed anywhere near the road. No rocks worthy of the name occur anywhere on the floor of the Sacramento Valley except around Sutter Buttes; any that may be needed for construction or landscaping must be hauled in from the mountains.

5

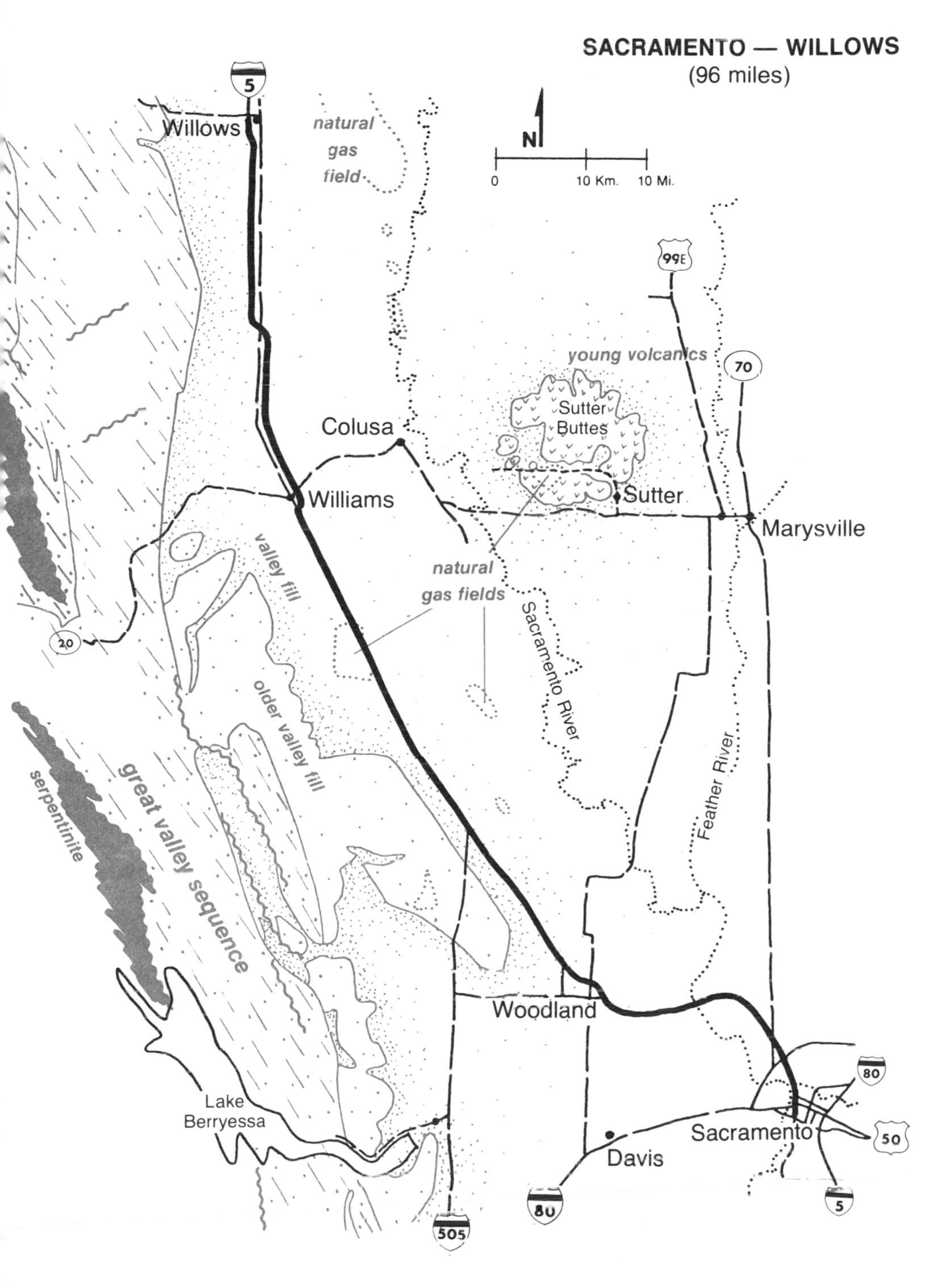
SACRAMENTO — WILLOWS
(96 miles)
N
0
10 Km.
10 Mi.
5
Willows
natural
gas
field
99E
young volcanics
70
Sutter
Buttes
Colusa
Williams
Sutter
Marysville
valley fill
natural
gas fields
Sacramento River
20
older valley fill
Feather River
great valley sequence
serpentinite
Woodland
Lake
Berryessa
80
Sacramento
50
Davis
80
5
505

SACRAMENTO — WILLOWS

Sutter Buttes, a defunct volcano about 20 miles northwest of Marysville, protrudes from the flat floor of the southern Sacramento Valley like a blister on a new paint job. This is the outstanding topographic feature, indeed the only one worth mentioning, in this part of the Sacramento Valley. Rising to an elevation of 2,132 feet, they make a striking landmark conspicuous for many miles.

Sutter Buttes have a long and rather complex history of volcanic activity dating from about 1½ million years ago. Curiously, this appears to be an isolated volcanic center not clearly related to any other. It is in line with the Cascades, but too far south to be considered one of them. But the rocks are similar to these erupted from the Cascade volcanoes and the age is about right so there must be some sort of connection. No one seems to know what it may be.

The first stage in development of Sutter Buttes was quiet intrusion of a large mass of molten igneous rock at shallow depth within the muddy sediments that had already filled this part of the valley. This must have blistered a large hill at the surface that promptly became the target of erosional processes which soon exposed the originally buried igneous rocks. Next there was intrusion of plug domes and volcanic eruptions, probably mostly steam explosions resulting from penetration of surface waters into extremely hot rocks buried within a few thousand feet of the surface. A series of

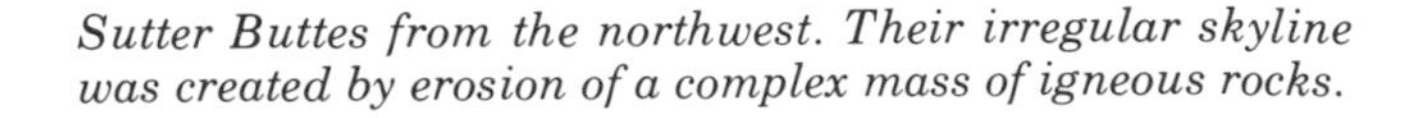
Sutter Buttes from the northwest. Their irregular skyline was created by erosion of a complex mass of igneous rocks.

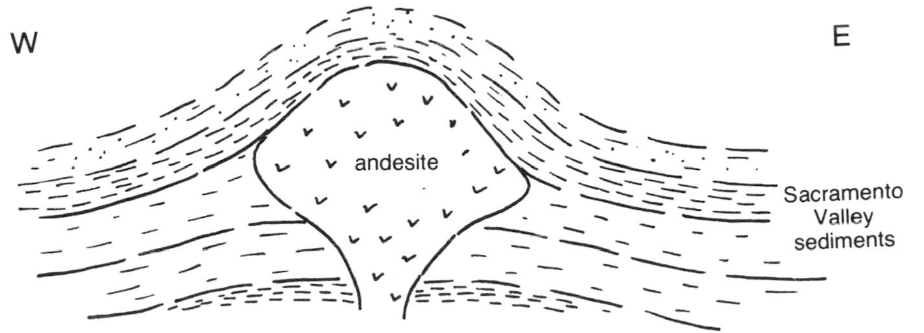

First stage: intrusion of igneous rocks — andesite

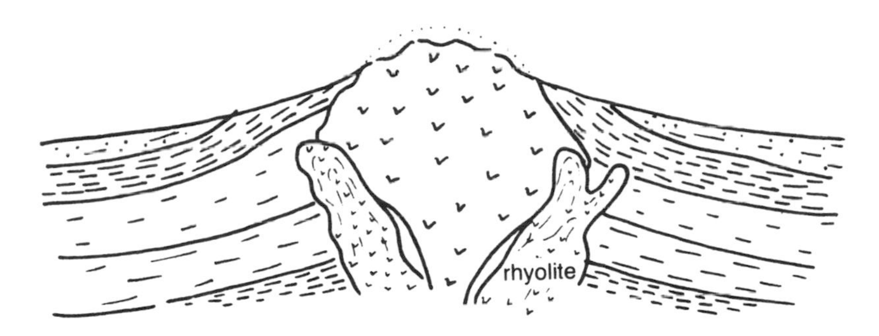

Erosion and intrusion of rhyolite plugs

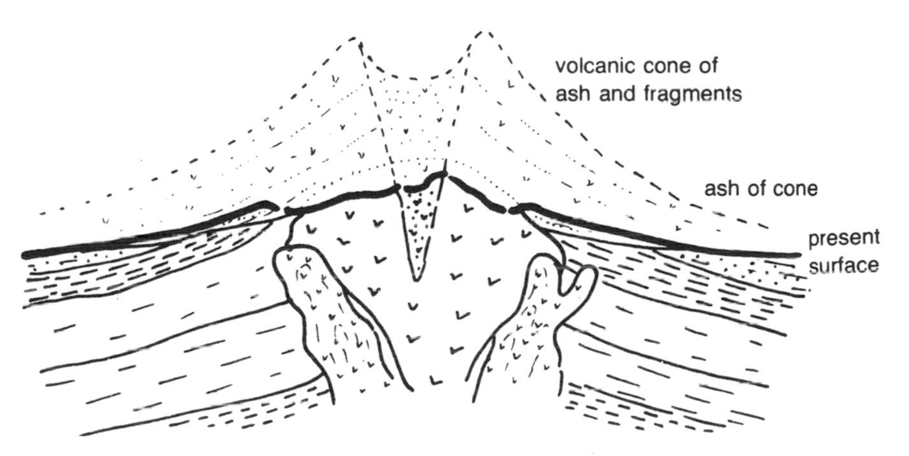

Volcanic activity and steam explosions

The three main stages in the long and complex volcanic history of Sutter Buttes.

Natural gas wells producing from layers of sedimentary rock folded against the flank of Sutter Buttes, background.

eruptions built a volcano composed mostly of shattered fragments of the original intrusive rock. Then the volcano was attacked by erosion and mostly removed, leaving the distinctive landscape of steeply hummocky remnants we see today.

An interesting side effect of the igneous activity at Sutter Buttes is the occurrence of large natural gas fields on its flanks. Sedimentary layers beneath the surface were arched and broken by faults adjacent to the igneous rocks to form traps that later caught large quantities of natural gas as it migrated along layers of porous sandstone. Natural gas, being very light, always tends to move upward until it comes to a place where the rocks are folded or broken in such a way as to block its progress.

5

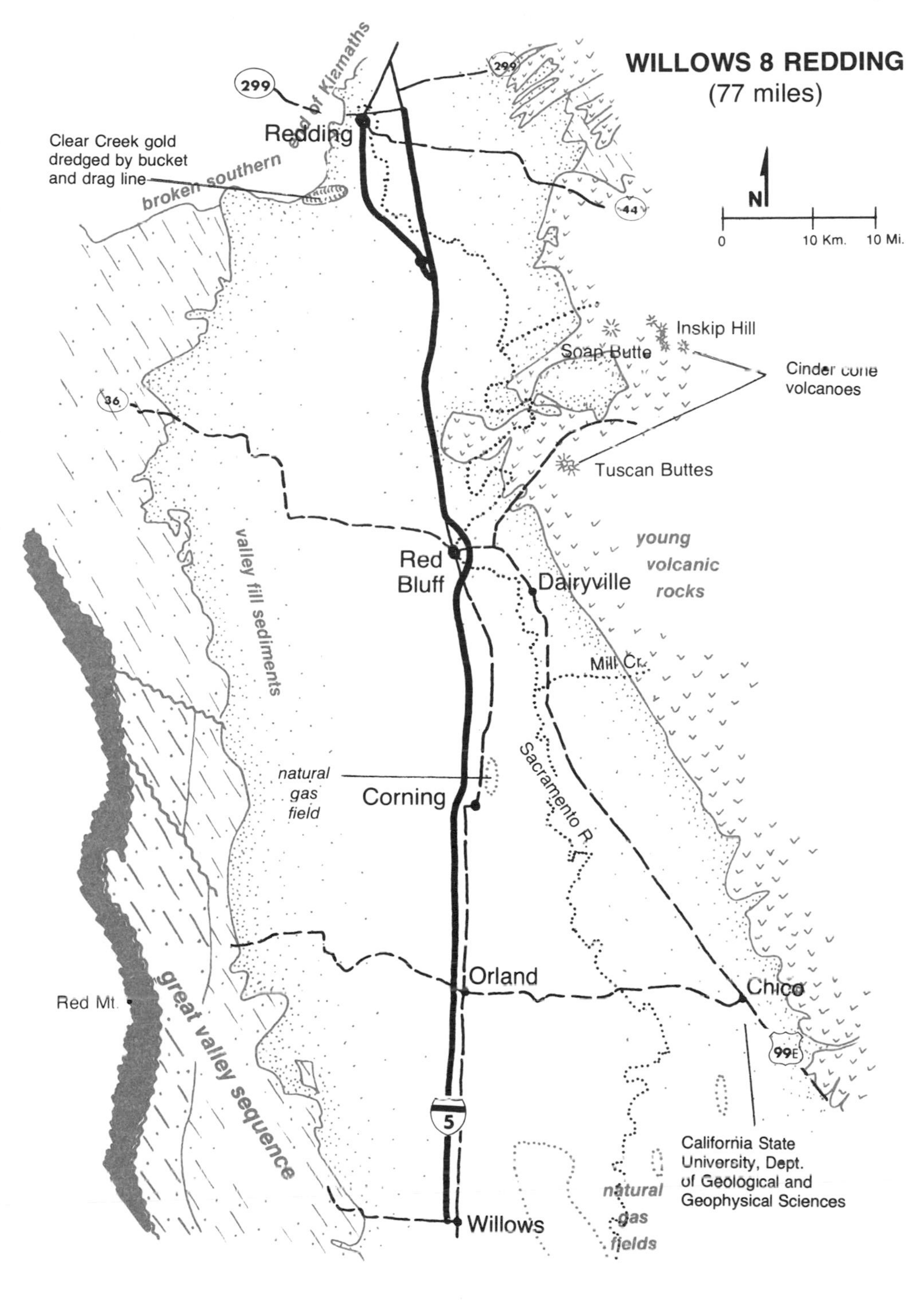
WILLOWS 8 REDDING
(77 miles)
N
0
10 Km.
10 Mi.
299
299
44
Redding
broken southern end of Klamaths
Clear Creek gold dredged by bucket and drag line
Inskip Hill
Soap Butte
Cinder cone volcanoes
36
Tuscan Buttes
young volcanic rocks
Red Bluff
Dairyville
valley fill sediments
Mill Cr.
natural gas field
Corning
Sacramento R.
Orland
Chico
Red Mt.
great valley sequence
99E
5
California State University, Dept. of Geological and Geophysical Sciences
natural gas fields
Willows

WILLOWS — REDDING

Between Willows and Red Bluff, Interstate 5 follows a route near the western side of the Sacramento Valley, crossing deposits of sediment — sand, mud, and gravel — washed in from the surrounding mountains within the last few million years.

These sediments were originally deposited as large alluvial fans, their gently-sloping surfaces spreading radially outward from the canyon mouths of the larger streams draining eastward from the Coast Range. Alluvial fans usually form in areas having very muddy streams, usually arid regions in which there are very few plants to protect the soil from erosion during occasional heavy rainstorms. Evidently these fans date from a time when the climate here was dry. More recently these same streams ceased to build their fans, and cut deep channels down through them, probably because their watersheds are now densely covered by plants which greatly reduce the muddiness of the streams. So the valley floor is not quite flat. The roadway gently rises and falls as it passes over the flanks of the old alluvial fans and down through the broad stream valleys now cut into them. Even though the old alluvial fans are too large and too dissected by later erosion to be seen easily from the ground, they reveal their presence in the subtly undulating topography of the valley floor.

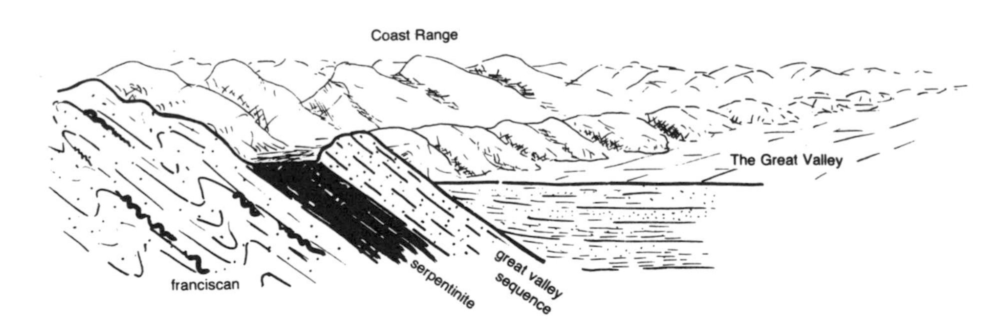

Rocks of the Coast Range are intricately crumpled beneath the Great Valley sequence and the muddy sediments of the Great Valley were deposited on top.

For long stretches of its route through the northern part of the Sacramento Valley, Interstate 5 closely parallels a long ridge which comes to a sharp crest about 10 to 15 miles west of the highway. It is continuous for long distances and its smooth slopes are nearly treeless, a sharp contrast to the darkly forested and intricately eroded slopes of the Coast Range clearly visible a few miles farther west. This long ridge is the outcrop of the thick layers of Cretaceous sandstones that geologists call the Great Valley sequence. It is a slice of nearly intact and undisturbed sea floor that rode above similar sea floor being stuffed under the margin of the continent to make the Franciscan rocks of the Coast Range. Many of the side roads that lead west from Interstate 5 pass this ridge through excellent roadcuts providing views of the rather monotonous brown sandstone beds of the Great Valley sequence, once the floor of the Pacific Ocean.

Looking across the flat floor of the Sacramento Valley toward the long ridge of Great Valley sequence sandstones that forms its western margin.

Between Red Bluff and Redding the floor of the Sacramento Valley becomes distinctly more rolling and tree-covered. High peaks of the Klamath Mountains are visible directly to the north and the Cascades seem to crowd in from the east. Here, obviously, is the northern end of the valley that seems in other places to go on forever.

Near Red Bluff, Interstate 5 follows the broad floodplain of the Sacramento River for several miles providing good views of the river itself. Early agriculture in the valley was mostly along this floodplain where irrigation water was easily obtainable. Like most streams that meander on broad floodplains, the Sacramento has natural levees along its banks, deposits of sediment laid down when muddy floodwaters rise onto the floodplain and dump part of their sediment load as they spread beyond the brushy stream banks. Natural levees are always difficult to see because they are normally no more than a few feet high, sometimes only a few inches, and may be several hundred feet wide. But they are always very important because they control patterns of drainage on the floodplain.

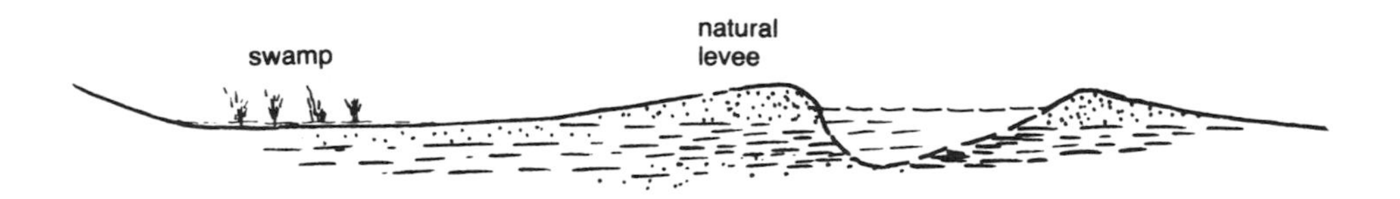

Natural levees form low ridges on the floodplain beside both banks of most meandering streams. Sometimes they get high enough to raise the stream above its floodplain.

Placer gold miners of the last century operating in the high mountain streams of the Klamaths and northern Sierra Nevada dumped large quantities of mud and sand into its headwaters, causing the Sacramento to become much muddier than it had been. The river responded to this insult, as rivers will, by depositing

more sediment than previously in its bed and on its natural levees thus raising the level of the stream slightly above large areas of its floodplain. Since water always runs downhill, the direct result was flooding large areas of floodplain converting once productive farmlands into extensive marshes and swamps more suitable for waterfowl and catfish. Thus miners in the remote streams of the high mountains ruined farmland in thevalley leading to a conflict of interests very similar to some of the modern environmental disputes. A California Supreme Court decision of 1884 prohibited further dumping of mine tailings into streams, an action that ended most hydraulic placer mining of gold in California. Today the river has largely repaired the damage, now that it again carries something like its normal sediment load, and most of the formerly marshy floodplain has been restored to agricultural use.

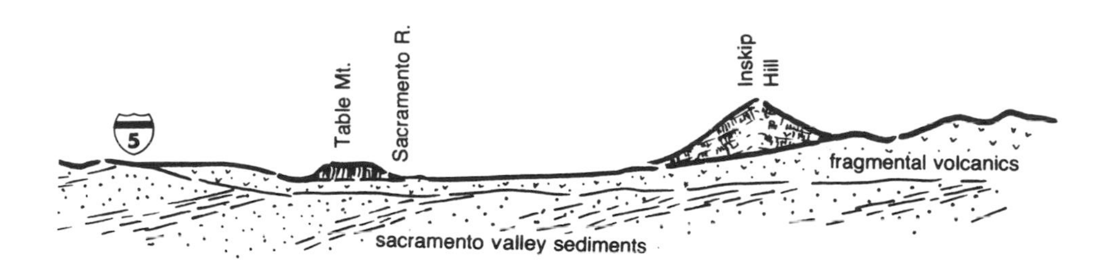

Section east from I-5 between Red Bluff and Redding. Cascade volcanic rocks overlap sedimentary fill in the Great Valley.

Extensive sheets of volcanic rock, mostly ash deposits, cover large parts of the surface of the Sacramento Valley east of Red Bluff and several small volcanic cones are visible from the interstate highway. This is the westernmost edge of the Cascades which begin almost directly due east of Red Bluff where the Sierra Nevada abruptly ends. The mountains directly north of Redding, the southern edge of the Klamaths, are the continuation of the Sierra Nevada, offset about 60 miles by a movement of the earth's crust.

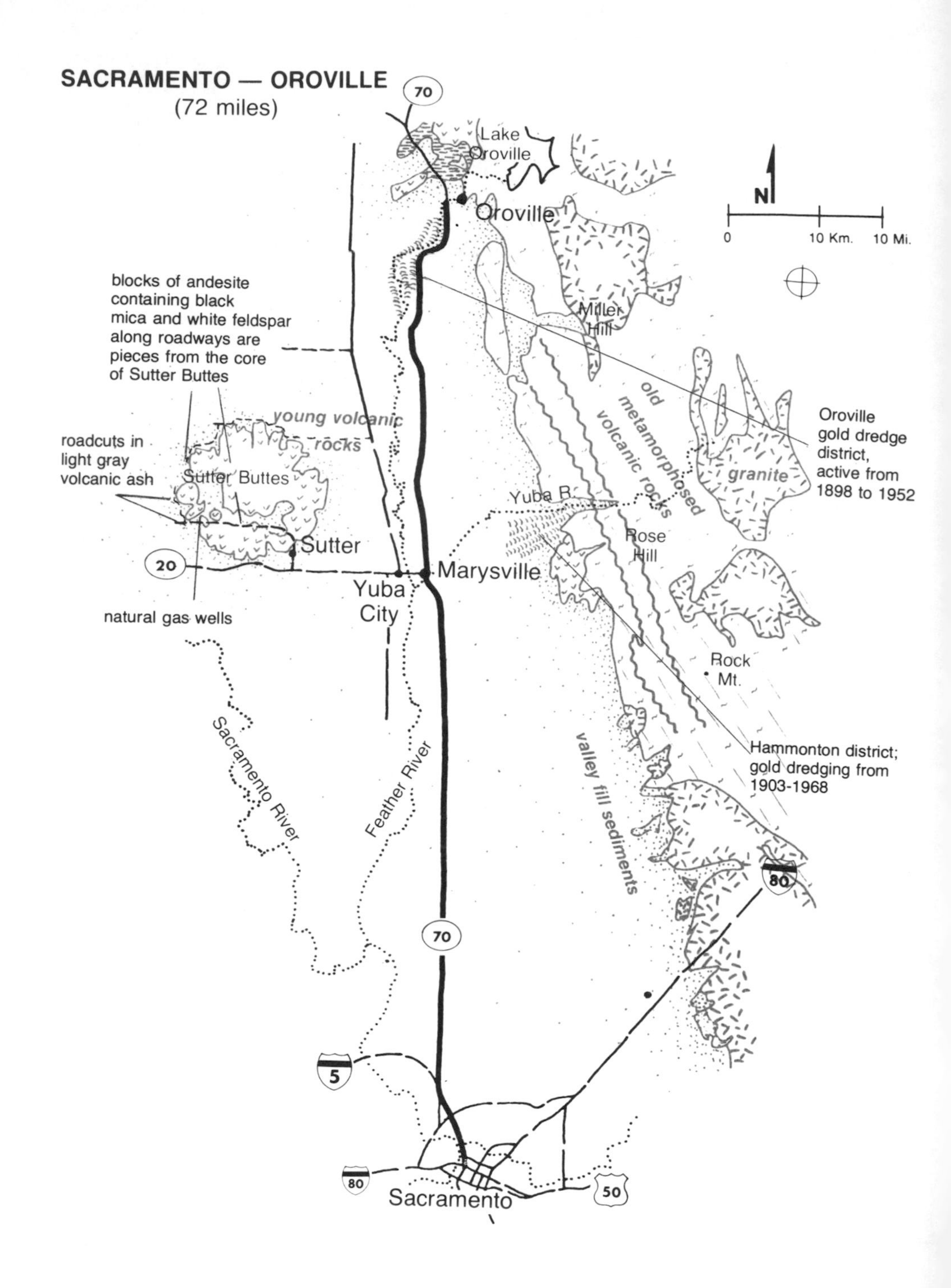
SACRAMENTO — OROVILLE
(72 miles)
70
Lake Oroville
Oroville
N
0
10 Km.
10 Mi.
blocks of andesite containing black mica and white feldspar along roadways are pieces from the core of Sutter Buttes
Miller Hill
old metamorphosed volcanic rocks
young volcanic rocks
roadcuts in light gray volcanic ash
Sutter Buttes
granite
Oroville gold dredge district, active from 1898 to 1952
Yuba R.
Rose Hill
Sutter
20
Marysville
Yuba City
natural gas wells
Rock Mt.
Sacramento River
Feather River
valley fill sediments
Hammonton district; gold dredging from 1903-1968
80
70
5
80
50
Sacramento

california 70
sacramento — oroville

California 70 follows the unbelievably flat floor of the Sacramento Valley all the way between Sacramento and Oroville. Softly rolling western foothills of the Sierra Nevada rise gently from the valley floor a few miles to the east and the distant ridges of the Coast Range form a low skyline in the west on days when the air is clear.

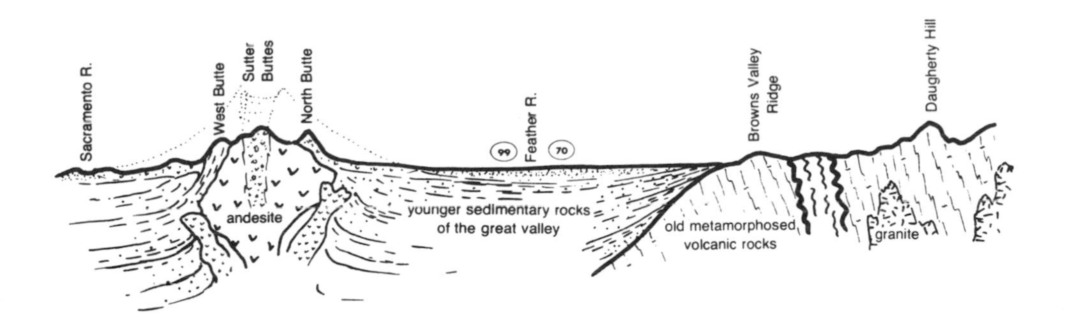

Section across the line of California 70 a few miles north of Marysville.

Just northwest of Marysville the flatness of the valley floor is punctuated by the Sutter Buttes, an isolated volcanic center that was active about 5 million years ago. Microscopic studies of the ash deposits around Sutter Buttes show that they are composed of tiny fragments of rocks that had already solidified before they were erupted. Evidently the eruptions were steam explosions that happened after the igneous rocks had solidified but while they were still very hot. Rocks of all kinds are excellent insulators and lose heat very slowly. It is common to find that a large body of molten magma buried beneath the surface may require millions of years to cool enough to solidify and more millions of years after that to actually become cool. So steam explosions can occur around volcanoes long after all other kinds of activity have ceased.

Dredges leave their spoils in neat rows which seem likely to remain for thousands of years before natural processes restore the land to productivity.

Between East Biggs and Oroville the road crosses a vast desolation of heaped piles of gravel left as spoils by the placer miners of the last century. This was the fertile floodplain of the Feather River until the gold miners found rich pickings in its gravels and worked them over with big dredges. The cost of restoring this land to its original productivity would far exceed the value of the gold that was mined from it and so eventually will the cost of lost agricultural production if it is not restored. Those who enriched themselves here left an enduring legacy of poverty to future generations.

CROSS REFERENCES

More about the Great Valley Sequence in the introduction to Chapter II, The Coast Ranges.

More about placer gold mining in Chapter III, U.S. 50: Sacramento — Carson City and California 49: Placerville — Vinton.

More about gold dredging in Chapter III, Interstate 80: Sacramento — Reno.

V

the cascades and modoc plateau

— a land of volcanoes

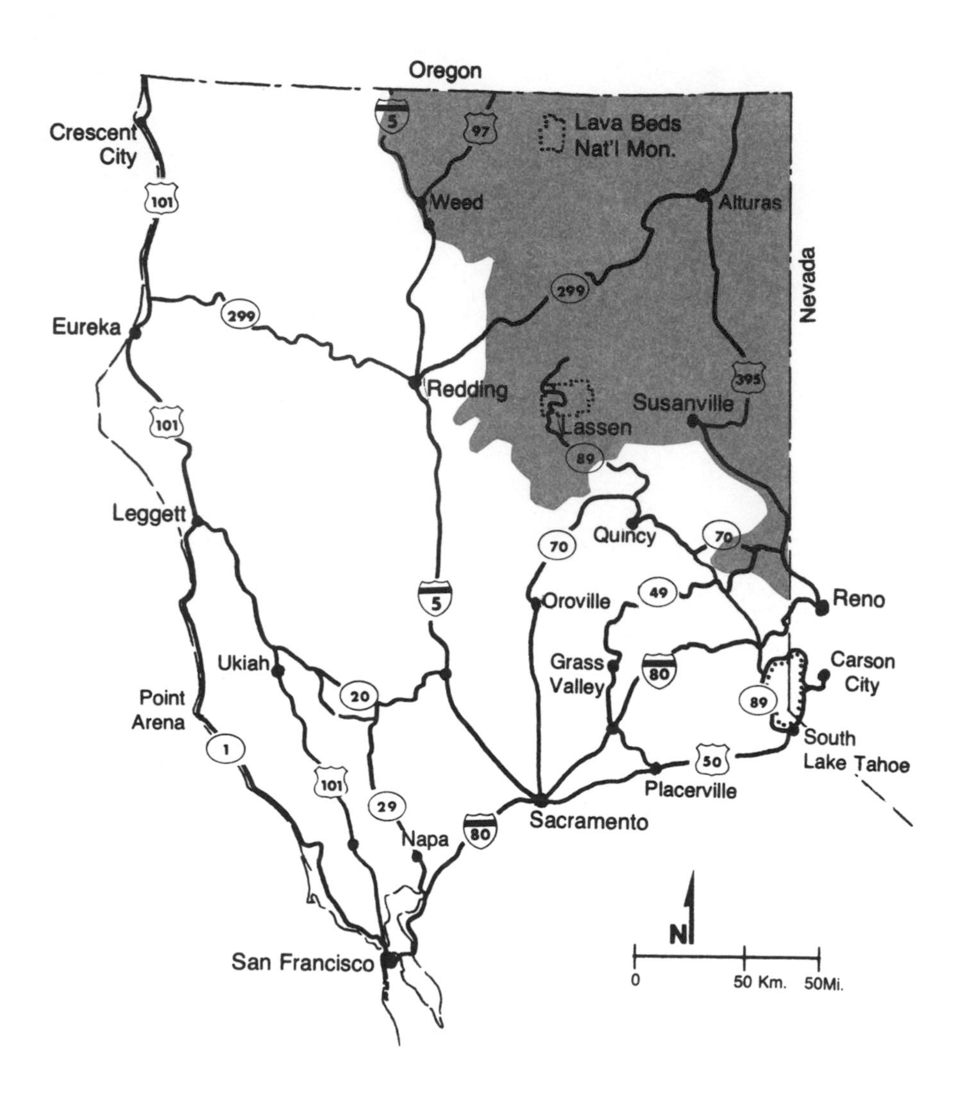

Oregon
Crescent City
Lava Beds Nat'l Mon.
Weed
Alturas
Nevada
Eureka
Redding
Susanville
Lassen
Leggett
Quincy
Oroville
Reno
Ukiah
Grass Valley
Carson City
Point Arena
South Lake Tahoe
Placerville
Sacramento
Napa
San Francisco
N
0
50 Km.
50Mi.

Rough surface of a young basalt flow looks like a river of black cinders. Rocks in foreground are older basalt thinly cloaked in vegetation.

Bubbly basalts cracked into crude vertical columns, a common sight in northeastern California

— a land of volcanoes

Once a seaway, then a plateau, now a region broken by faulting into mountains and bordered on the west by a wall of volcanoes, California's volcanic northeast is the southernmost tip of a province that includes much of the Pacific Northwest. Except for local deposits of muddy sediment washed into valley floors and lakes, nearly every rock in this entire region was erupted as molten lava sometime within the last 30 million years.

Part of the space in the continent for California's Cascades and Modoc Plateau was opened about 140 million years ago when the Klamath and Sierra Nevada blocks separated and moved about 60 miles apart to their present positions. Most of whatever rocks may record the next events in the area newly embraced by the Klamaths to the west and the truncated northern end of the Sierra Nevada to the south are now buried under younger volcanic rocks. But outcrops exposed along the eastern flank of the Klamaths including some crossed by Interstate 5 between Yreka and the Oregon line provide a tantalizing glimpse into the pre-volcanic past. The rocks consist of muds and sands containing fossils of animals that lived in the oceans during late Cretaceous time — until about 60 million years ago. Evidently northeastern California, like the Great Valley was flooded by seawater then and the Klamath Mountains must have been an island, or nearly so.

Large quantities of natural gas are produced from late Cretaceous sedimentary rocks in the Sacramento Valley generally similar to those exposed here along the east flank of the Klamath Mountains. If the beds of rock on the eastern flank of the Klamaths extend eastward beneath the younger volcanic landscape of the Modoc Plateau, as seems very likely, then new reserves of natural gas and petroleum may await discovery in northeastern California.

The seaway filled with sediment and remained as a level plain similar to the present Great Valley for millions of years. Then, incandescent floods of molten basalt welled up from long fissures and poured fluidly across the level sediments to form enormous lava flows covering hundreds of square miles to depths of as much as several hundred feet. Beginning about 30 million years ago, these eruptions continued for a period of about 15 million years and built a high lava plateau covering much of the Pacific Northwest. A plateau formed instead of mountains because the lava spread thinly across the countryside from long fissures, instead of piling up around the vent as do more viscous lavas. Iceland is the only country where such eruptions have occurred during historic time; there have been some in southern Idaho within the last few thousand years.

Basalt is the commonest volcanic rock and the easiest to recognize. Fresh and unweathered, it is always black, an especially grim and uncompromising flat black, generally without glossiness or color. Old surfaces often become barn red as weathering stains them with iron oxides, and some basalts acquire a greenish cast to their blackness through alteration by water. Outcrops of basalt always have a grimly forbidding look darkly suggestive of this rock's origin deep beneath the continental crust.

Basalt lavas rise from beneath the continental crust, from depths of about 100 miles or more where the rocks are red hot and retain a tenuous hold on their solidity only because they exist under tremendous pressure. Where temperatures rise higher, or the pressure is somewhat relieved, melting begins with basalt as the first fluid to form. Basalts erupt on continents in places where the crust is stretched under tension, a situation that eases pressure on rocks beneath the continent, and they erupt on the ocean floors at mid-ocean ridges where hot "convection currents" rise and spread the surface. It is difficult to know which mechanism produced the floods of basalt that inundated much of the Pacific Northwest between 30 and 15 million years ago. Perhaps the East Pacific Rise, a rising "convection cell," was active beneath the western part of our continent during the time when the basalt floods poured onto its surface.

Basalt lava flows, the commonest volcanic rock, are generally the most predictable of all rocks. They look about the same whether they were erupted from volcanoes in California, Hawaii or the Moon. Modoc Plateau basalts are a little different; they contain far more gas bubbles than most of those elsewhere and often lack the distinctive tendency of basalts to break into rows of neat vertical columns, like stockades of fenceposts, where they are exposed in cross-section. Most basalts break into such geometrically regular vertical columns along fractures that developed in the rock as it shrank upon cooling. Bubbly basalts such as those on the Modoc Plateau seem able to distribute shrinkage stresses uniformly through the rock without developing a regular pattern of fractures.

Gas bubbles in basalts form as the molten magma releases steam after it is erupted. Evidently the magmas erupted to form the Modoc Plateau must have picked up an extra charge of steam somewhere along the way from their source in the mantle to the earth's surface. Perhaps they took on water as they rose through the deep accumulation of watery muds filling the seaway that once existed here.

After the major flood basalt eruptions ceased in northeastern California, the earth's crust under much of western North America continued to stretch, and broke into large blocks that moved vertically to form mountain ranges and basins. A vast region was broken, extending from Colorado to the Sierra Nevada and from Montana to central Mexico. Parts of the lava plateau in northeastern California and southeastern Oregon were included and broke into block mountains and basins.

Mountain and basin blocks in western North America are elongated in a north-south direction, a fact that prompted an early geologist to remark that their map pattern suggests an army of caterpillars marching north out of Mexico. Modern geologists suspect that this pattern indicates that the blocks owe their origin to crustal stretching in an east-west direction. Certainly they were created by movement of the rocks beneath the continental crust, perhaps by spreading away from the crest of a spreading zone beneath the continent.

Formation of block fault mountains is always accompanied by eruption of basalt flows and light-colored volcanic ash, called rhyolite. Rhyolite, unlike basalt, has a composition similar to granite and almost certainly forms by partial melting of rocks within the continental crust. Both kinds of lava are melted by input of extra heat from the earth's interior aided by relief of pressure at depth as the stretching crust cracks and thins.

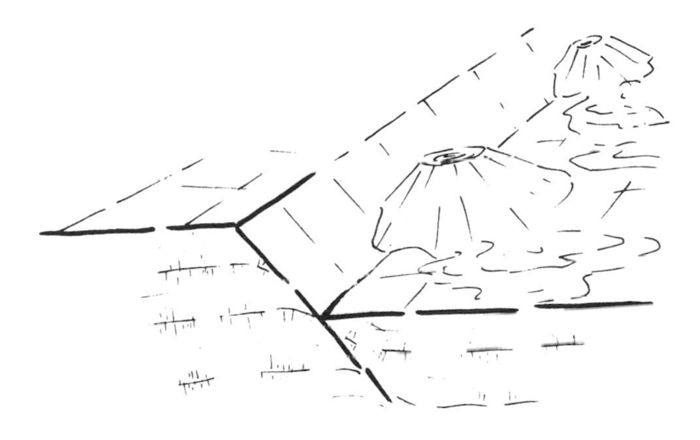

Eruption of these later basalts was quite different from that of the earlier plateau basalts, probably because the lava was not produced in such overwhelming quantity. Modest amounts of basalt emerged from single vents, instead of from fissures many miles long, and produced rather small lava flows that generally cover areas of less than 10 square miles. Shreds of lava coughed out of the vent by escaping steam piled up to make small cinder cone volcanoes a few hundred feet high.

Unlike basalt, rhyolite lava is extremely viscous, generally about the consistency of putty and is often heavily charged with steam. Violently escaping steam often blasts the lava out of the volcano as a glowing spray of red hot shreds of molten rock that cool as they drift downwind and blanket the countryside in light-colored volcanic ash. Sometimes the droplets settle still molten and then weld themselves together to encase the ground surface in an instant armor of solid rock. Sometimes, apparently when they are nearly free of steam, rhyolite lavas ooze quietly out of the vent to form thick, swollen-looking, lava flows that cool to make the glossy black volcanic glass called obsidian. The area south of Lava Beds National Monument contains several of the finest obsidian flows in the country.

Volcanic rocks old enough to be dissected and exposed by erosion, such as those on the Modoc Plateau, are often sources of a variety of semi-precious gemstones. Petrified wood, agate and opal, all forms of the common mineral quartz, are not volcanic even though they often occur in volcanic rocks. Trees buried in volcanic ash become petrified in time and most rhyolitic ash beds contain at least some petrified wood, it is usually easiest to find as pebbles in stream gravels. Agate and opal form as circulating ground water dissolves quartz out of the volcanic rocks and redeposits it in open pore spaces.

Construction aggregate and road metal are the major commodities mined from volcanic rocks in the Modoc Plateau. Basalt lava flows rarely contain valuable mineral deposits of any kind and those in northeastern California are not exceptions. Rhyolitic volcanic rocks frequently contain deposits of valuable minerals, most often ores of silver, tin, and copper but occasionally others. However, no major deposits are known in the rhyolites of the Modoc Plateau.

Another kind of economic resource characteristically abundant in areas of fairly recent volcanic activity is natural steam, useful for generating electrical power. Development of natural steam resources will undoubtedly be much more important in the future than ever in the past, and volcanic provinces such as the Modoc Plateau are obvious places to prospect for steam.

Northeastern California's most recent volcanic episode has been development of the Cascade volcanoes, a continuing process that began several million years ago. A wide variety of lavas, including basalt and rhyolite, are erupted from the Cascade volcanoes but the commonest are andesites, rocks intermediate in composition between rhyolite and basalt. Andesites vary considerably in appearance but most are some shade of gray or brown. Like all volcanic rocks, they are so very fine-grained that it is difficult to see anything very distinctive in them.

Andesites are as variable in their eruptive behavior as in their color, generating both lava flows and ash falls, sometimes quietly and other times violently. They frequently build picturesque, symmetrical cones composed of ash deposits and volcanic mud flows, tied into a reasonably solid mass by lava flows. These are the high, snowcapped cones so much admired for their beauty — Mt. Shasta among others.

Andesitic volcanoes typically form long, curving chains parallel to the seacoast. Usually they occur immediately continentward of a deep trench in the sea floor and are generally associated with busy zones of earthquake activity. Most geologists are now convinced that such volcanic chains appear where the descending arm of a convection current is carrying sea floor deep into the hot interior of the earth. Lavas erupted from Cascade volcanoes are probably material making up the floor of the Pacific Ocean until it was swept down into the earth's interior and melted several million years ago. There have been very few eruptions from Cascade volcanoes in recent centuries and the area does not have many earthquakes today. Apparently the action has either stopped or changed in character during the very recent past.

Volcanic processes are directly expressed in the landscape of the Cascades which is virtually unaffected by either erosion or faulting. There are very few streams in either the Cascades or the Modoc Plateau because the young volcanic rocks are still so porous that surface water soaks into the ground instead of running off.

Recently erupted andesites, like those in the Cascades, rarely contain mineable deposits of valuable minerals. But these rocks are still hot and certain to contain natural steam suitable for generating electricity.

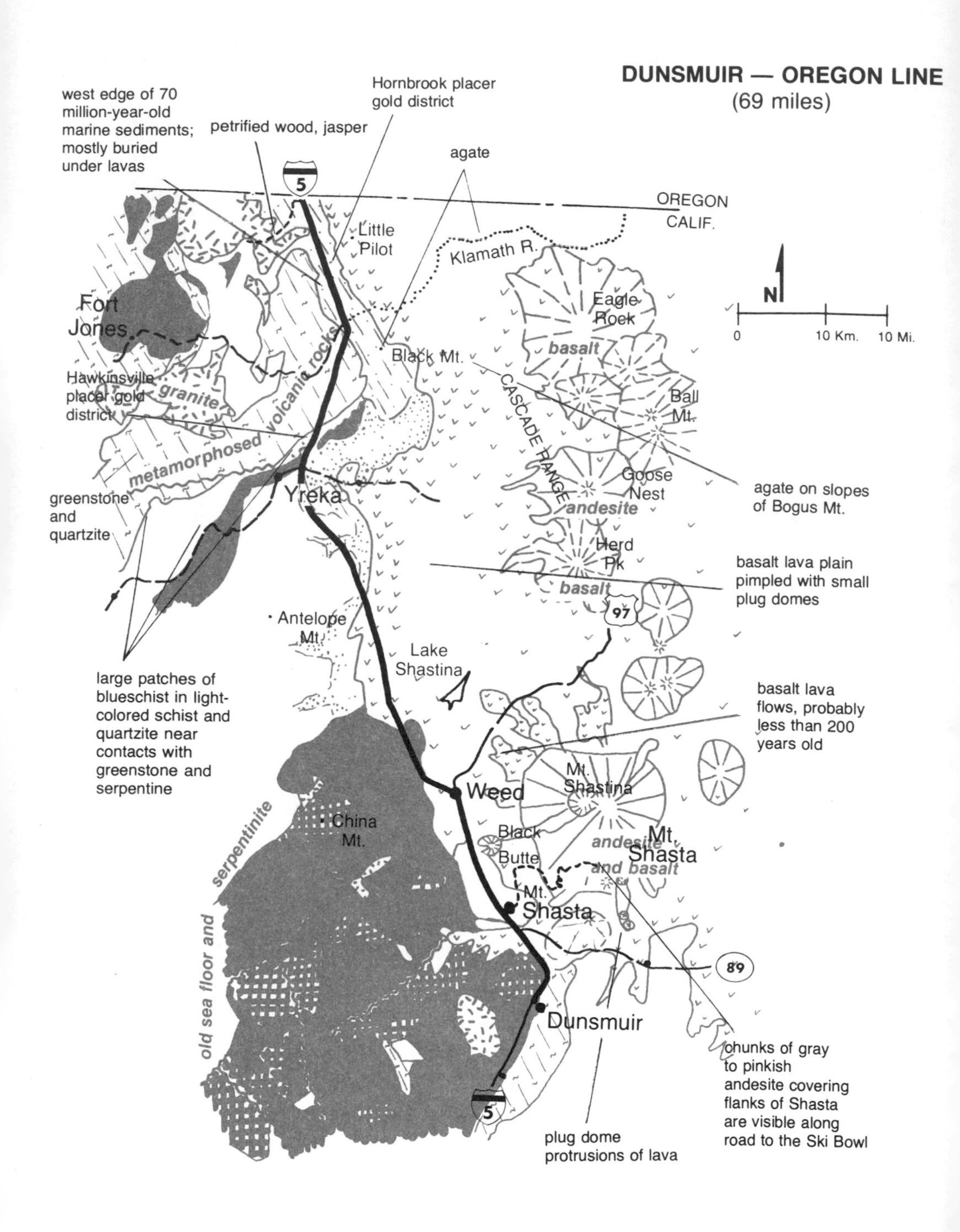

DUNSMUIR — OREGON LINE
(69 miles)
west edge of 70 million-year-old marine sediments; mostly buried under lavas
petrified wood, jasper
Hornbrook placer gold district
agate
OREGON
CALIF.
Little Pilot
Klamath R.
N
0
10 Km.
10 Mi.
Eagle Rock
basalt
Fort Jones
Black Mt.
Hawkinsville placer gold district
granite
volcanic rocks
metamorphosed
CASCADE RANGE
Ball Mt.
agate on slopes of Bogus Mt.
greenstone and quartzite
Yreka
Goose Nest
andesite
Herd Pk
basalt
basalt lava plain pimpled with small plug domes
97
Antelope Mt.
Lake Shastina
large patches of blueschist in light-colored schist and quartzite near contacts with greenstone and serpentine
basalt lava flows, probably less than 200 years old
Mt. Shastina
Weed
China Mt.
serpentinite
old sea floor and
Black Butte
andesite and basalt
Mt. Shasta
Mt. Shasta
89
Dunsmuir
chunks of gray to pinkish andesite covering flanks of Shasta are visible along road to the Ski Bowl
plug dome protrusions of lava

interstate 5
dunsmuir — oregon line

Between Dunsmuir and Weed, Interstate 5 crosses volcanic rocks on the westernmost edge of the Cascades where they lap onto the Klamath Mountains. Volcanoes form the landscape east of the road whereas the hills to the west are eroded into much older rocks of the Klamaths.

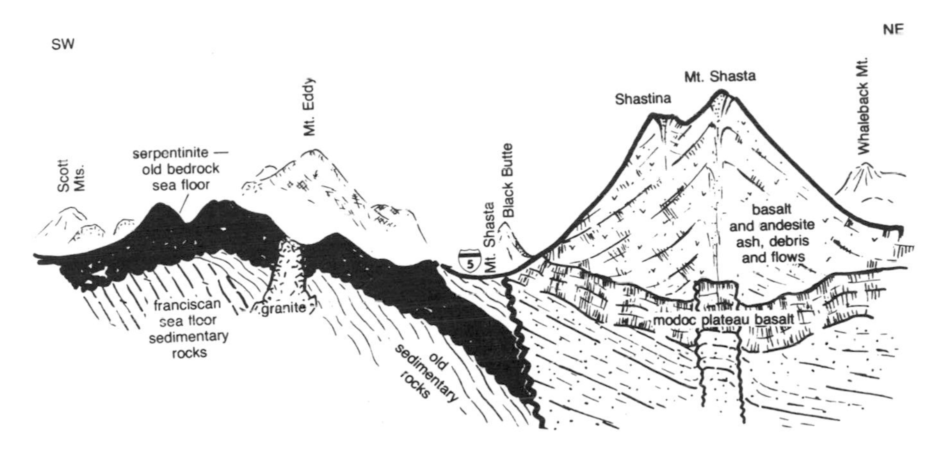

Section across I-5 at Mt. Shasta. The younger volcanic rocks of the Cascades lap onto the older rocks of the Klamaths to the west.

Shasta dominates the scene northeast of Interstate 5 all the way from Dunsmuir to Weed. Its massive volume of about 80 cubic miles was erupted during the last million or so years in outbursts that produce large numbers of lava flows, mostly composed of brown andesite and in its final stages of activity, ash falls. Like most volcanoes built by this type of eruption, Shasta has steep sides and a gracefully symmetrical form, deeply gouged in places by glacial valleys. Shastina, the large, smooth lava cone forming the west peak, was formed around a single vent since melting of the ice-age glaciers. The thin layer of brown pumice that mantles the mountain is from the latest eruption of Mt. Shasta, apparently in 1786.

Curving ridges of bouldery debris in the floor of this ice-scoured valley are glacial moraines. Mt. Shasta Ski Bowl.

About 5 miles south of Weed, Interstate 5 skirts the steep western flank of Black Butte, an example of an interesting type of craterless volcanic mountain called a plug dome. These are large masses of volcanic rock that erupt by squeezing slowly out of the earth over a period of months. Masses of viscous magma bulge upward, like enormous toadstools, shedding their outermost layers of cooling rock as sliding heaps of steaming rubble to form steep-sided mountains clothed in talus. Angular blocks of reddish-brown andesite cover the sides of Black Butte so completely that no solid rock is exposed except in a few craggy outcrops near the summit. Plug domes are rather common in the Cascades. Several others are visible from Interstate 5 as steep-sided knobs protruding, like giant warts, from the south slopes of Shasta.

Immediately west of the Shasta area, large expanses of the Klamath Mountains are composed of rocks that once formed the ocean floor. Coarse-grained black igneous rocks and fine-grained black basalt lava flows are still covered in a few places by ocean-bottom sediments. Numerous small intrusions of granite, Klamath equivalents of the Sierran batholiths, penetrate the black oceanic rocks and form some of the higher mountains.

Black Butte, a craterless plug dome cloaked in a loose rubble of reddish andesite.

Cross-section of Black Butte. Blocks cracked from the outer shell of the mass as it cooled clothe its slopes in rubble.

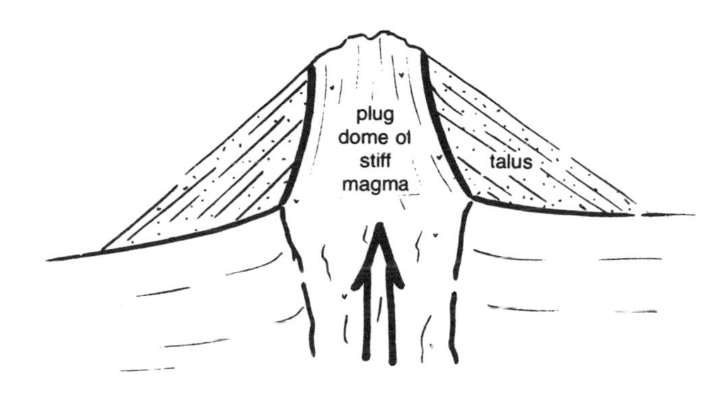

Between Weed and Grenada Junction, Interstate 5 closely follows the boundary between the volcanic Cascade province to the east and the Klamath Mountains to the west. All bedrock exposures along this route are volcanic rocks that were erupted during Miocene time, 10 to 25 million years ago, and then deeply carved by erosion before the present Cascade volcanoes even began to form. Much of the road is built on flood plain deposits laid down by streams flowing through valleys eroded into the volcanic rocks.

At Grenada Junction, Interstate 5 turns northwest leaving the volcanics behind as it enters the Klamath Mountains. Bedrock exposed in the hills on both sides of the valleys followed by the road is ancient sedimentary rock originally laid down on the floor of the Pacific Ocean during Ordovician time, approximately 475 million years ago. Black igneous rocks, the old sea floor on which these sediments were deposited, is in the hills east and west of Yreka. These are among the oldest rocks in northern California.

Interstate 5 generally follows the valleys of Yreka Creek and the Klamath River for a distance of about 11 miles north of Yreka. Bedrock along this stretch of road is mostly old volcanic and sedimentary rocks originally laid down on the ocean floor sometime during the Paleozoic Era, several hundred million years ago. Events since then have changed these rocks considerably as they have been folded, broken by faults, and cooked in the presence of hot water. Outcrops are few in these heavily forested mountain slopes deeply mantled by soil. Where they are visible, the rocks tend to be darkly greenish in overall color and rather characterless in general appearance.

From about a mile south of the bridge across the Klamath River to the Oregon line, Interstate 5 is built on Cretaceous sedimentary rocks, mostly greenish and gray sandstones and mudstones deposited about 70 million years ago in the seaway that once flooded the Modoc Plateau. Hills in the distance east of the road are underlain by much younger volcanic rocks, those west of the road by much older rocks belonging to the Klamath Mountains. This stretch of Interstate 5 closely follows the boundary between the two geological provinces.

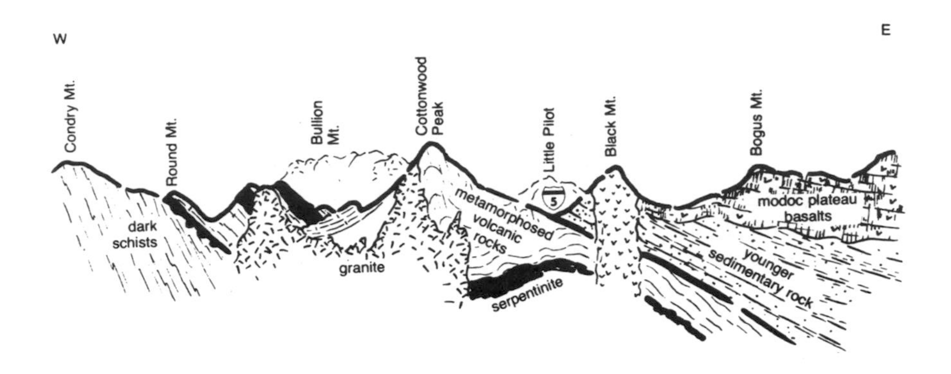

Section across I-5 a few miles south of the Oregon line. Granites intruded the Klamath Mountains while muddy sediments were deposited on their eastern flank. The volcanics are much younger.

Fossils found in the Cretaceous sandstones and mudstones indicate that these rocks were deposited in sea water. They are evidence that a seaway existed in northeastern California and filled with sediments, just as did the Great Valley. It was later covered by the volcanics exposed in the hills east of the road. This area along Interstate 5 is one of the very few places where these sedimentary deposits peek out from under the edge of the volcanic plateau that buries them. The layers appear to extend eastward beneath the Cascade volcanoes and the Modoc Plateau, and may contain petroleum and natural gas. It will be necessary to drill through thousands of feet of volcanic rocks to discover whether such deposits exist.

WEED — OREGON LINE
(55 miles)

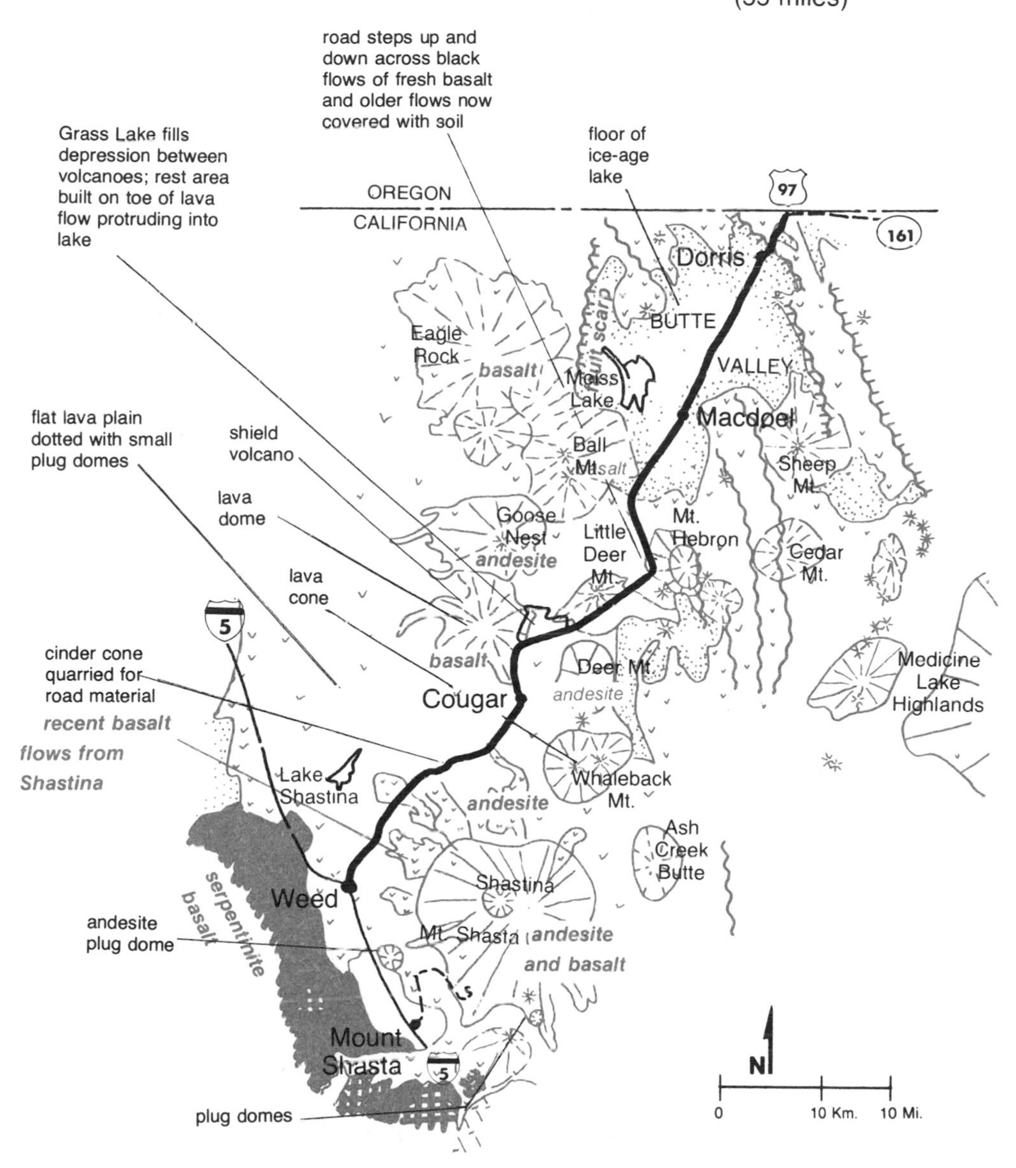

Craggy Mt. Shasta, left, has been glacially eroded. Shastina, right, is very young and unmarked by erosion.

u.s. 97
weed — oregon line

All the rocks exposed along this road are volcanic, none of them more than a few million years old and many no more than a few thousand. Most of the landscape is likewise volcanic having been created by the same eruptions that formed the rocks and still too young to be reshaped by the processes of erosion.

Weed is in the Shasta Valley, a trough between the Klamath Mountains to the west and the Cascade volcanoes to the east. Most of the valley floor is covered by volcanic rocks erupted about 20 million years ago. These were considerably carved by erosion before being mostly buried beneath the Cascade volcanoes during the past million or so years.

Section across U.S. 97 at Cougar. The sedimentary rocks were deposited while this area was a sea after separation of the Klamaths and Sierra Nevada. Younger volcanic rocks cap them.

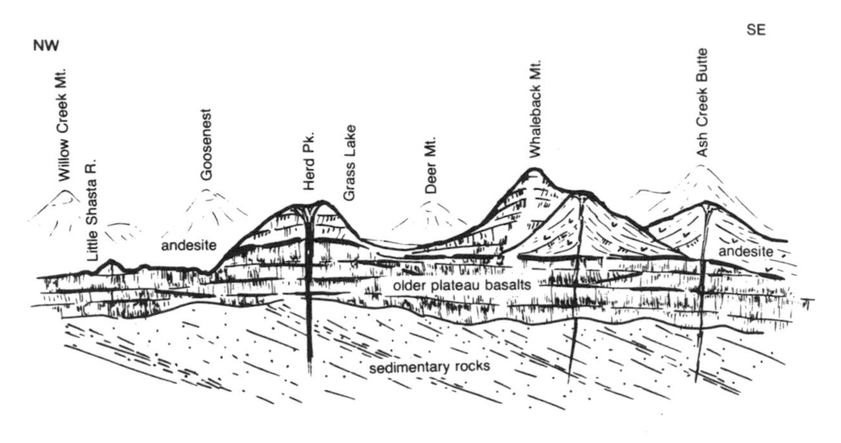

Brown volcanic ash blown from Shastina mantles the landscape north of Weed.

Shasta's snowcapped double peak broods gracefully over much of northernmost California and fills the skyline southeast of Weed. Like the other Cascade volcanoes, it is very young. Many of the rocks on its surface are only a few hundred years old and it may well be an active volcano even though it has not erupted lately. A large part of its bulk has been erupted within the last few thousand years — since the last ice age.

Shasta's main cone looks a bit ragged because it has been deeply gouged by glaciers, mostly since the last ice age, and has not been repaired by new eruptions. Several small glaciers exist on the mountain now. Some of those glaciers were large enough during the last ice age to reach the floor of the Shasta Valley and leave sizeable moraines in the area between Weed and the east side of Lake Shastina.

Shastina, the subsidiary cone on the west flank of Shasta, is perfectly fresh and unmarked by erosion; it must have formed within the last several hundred years and appears to have been the site of the most recent eruptions. The latest of these may well have occurred as recently as 1786 when a ship exploring the coast reported seeing an active volcano in this vicinity.

Between 5 and 10 miles north of Weed, U.S. 97 skirts the northern edge of a very large black lava flow erupted from Shastina within the last few hundred years. Its steep front covered with rubbly blocks of raw rock and nearly bare of vegetation is conspicuous a short distance south of the road.

Between Weed and Macdoel U.S. 97 crosses the Cascades, winding its way between volcanoes and over low lava flows. There are two abundant kinds of volcanic rock in this area: basalt, which is always absolutely black on fresh surfaces but with time often weathers to a barn-red, and andesite, which comes in various shades of gray or brown. Both rocks are mostly very fine-grained, and smooth looking, but may contain scattered crystals large enough to see and either may be full of holes made by bubbles of steam escaping from the lava while it was still molten. Basalt lava is a very runny liquid so it spreads over large areas to make thin lava flows that accumulate to form broad volcanoes with gently-sloping sides. Andesite lava is much thicker so it piles up around the vent to make steep-sided volcanoes. The shape of the volcano is a clue to the kind of rock it contains. Several of the volcanoes visible from U.S. 97 are capped by small, steep-sided cinder cones composed of volcanic rock fragments blown from the crater during its latest eruption.

About six miles southwest of Macdoel U.S. 97 leaves the Cascades, descending a steep slope onto the flat floor of Butte Valley, one of the dropped fault blocks of the Modoc Plateau. Older volcanic rocks, mostly basalt, that were erupted before development of the Cascade volcanoes, form the bedrock here.

As might be guessed from its very flat floor, crossed by U.S. 97 between Macdoel and Dorris, Butte Valley once contained a lake and is now underlain by lake sediments, mostly clays. The lake existed during the last ice age, when the climate was much wetter than it is today, and drained northward into the Klamath River. All that remains, now that the climate has been much drier for several thousand years, is a marshy area called Meiss Lake, that has no outlet and fluctuates in size from one year to another.

395

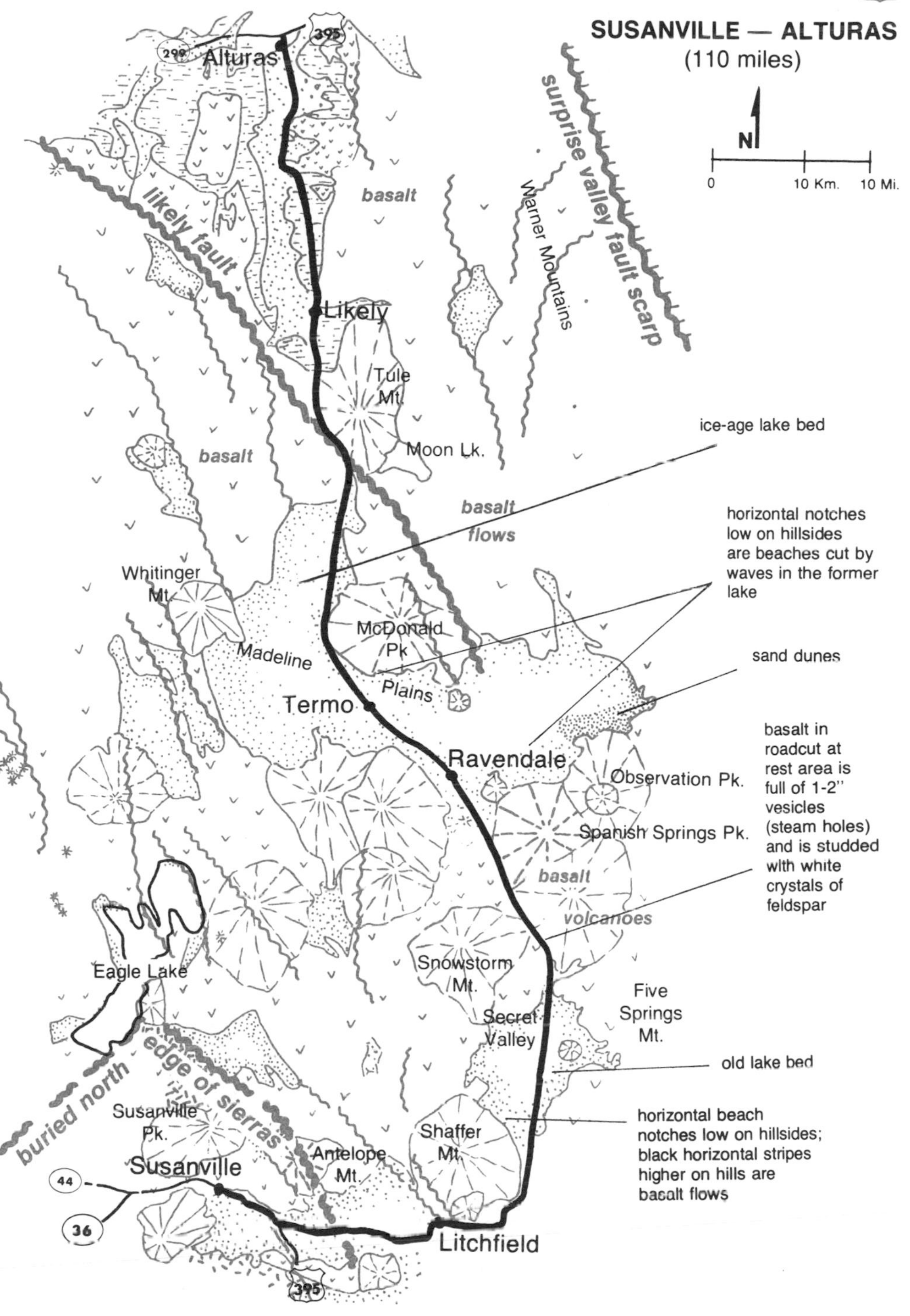
SUSANVILLE — ALTURAS
(110 miles)
N
0
10 Km.
10 Mi.
299
395
Alturas
basalt
likely fault
Likely
surprise valley fault scarp
Warner Mountains
Tule Mt.
Moon Lk.
basalt
ice-age lake bed
basalt flows
horizontal notches low on hillsides are beaches cut by waves in the former lake
Whitinger Mt.
McDonald Pk
Madeline
Plains
Termo
sand dunes
Ravendale
Observation Pk.
Spanish Springs Pk.
basalt in roadcut at rest area is full of 1-2" vesicles (steam holes) and is studded with white crystals of feldspar
basalt volcanoes
Eagle Lake
Snowstorm Mt.
Secret Valley
Five Springs Mt.
old lake bed
buried north edge of sierras
Susanville Pk.
Shaffer Mt.
horizontal beach notches low on hillsides; black horizontal stripes higher on hills are basalt flows
Antelope Mt.
Susanville
44
36
Litchfield
395

u.s. 395
susanville — alturas

Volcanic rocks, mostly lava flows of black basalt locally accompanied by deposits of white volcanic ash, form the bedrock along this road. Most of them were erupted during the last 10 million years while the earth's crust was breaking along faults into large blocks that moved up and down to form the present pattern of mountain ranges and basin valleys.

Higher mountains to the west cast their rain shadow over this area, giving it a very dry climate. Enough so that no large rivers drain the fault-block valleys, so they are slowly filling with sediment as their surrounding mountains erode away. All these undrained basins filled with water to become lakes while the climate was briefly much wetter during the ice ages. No doubt the hills were green then and the landscape must have been lush. Now the climate is dry again, the hills are thinly cloaked with scrubby vegetation, and all that remains of the lakes are shrunken remnants rapidly becoming salty and alkaline because they have no outlets.

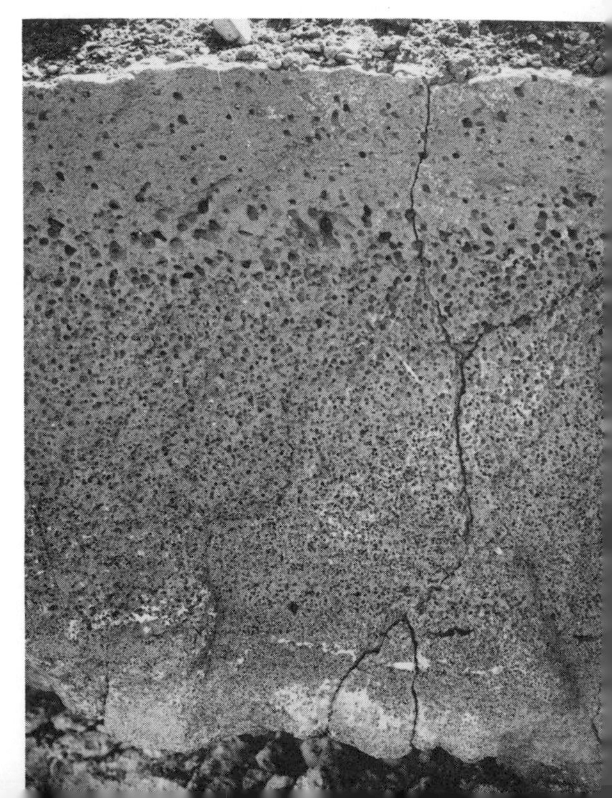

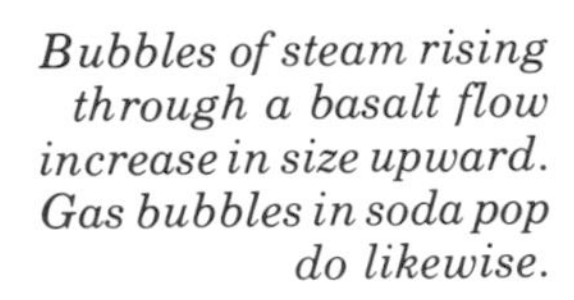

Bubbles of steam rising through a basalt flow increase in size upward. Gas bubbles in soda pop do likewise.

The shoreline of an ice-age lake is preserved as a notch in this hillside east of U.S. 395 South of Ravendale.

U.S. 395 follows close to what was once the north side of Honey Lake for nearly fifteen miles east of Susanville before turning north to head towards Alturas. Hills rising in the distance southwest of the lake basin are composed of granite; they are the northernmost outposts of the Sierra Nevada. After turning north, the road passes for 27 miles through rugged hills eroded into volcanic rocks and across Secret Valley, a small fault-block basin that also contained a lake, before coming down onto the flat surface of the Madeline Plains about 3 miles south of Ravendale.

Their extremely flat surface suggests that the Madeline Plains are underlain by lake bed deposits; the old lakeshore is still clearly visible as a perfectly horizontal line on the hillside.

North of the Madeline Plains, U.S. 395 crosses another 10-mile stretch of rugged hills eroded into basalt lava flows and then descends into a long, straight valley in which it follows the South Fork of the Pit River from Likely north to Alturas. Most of this stretch of road is built across volcanic and sedimentary rocks which partially filled the basin created when a large block of the earth's crust dropped along faults several million years ago. The Warner Mountains, visible on the skyline 25 miles to the east, are another block that was raised at about the same time.

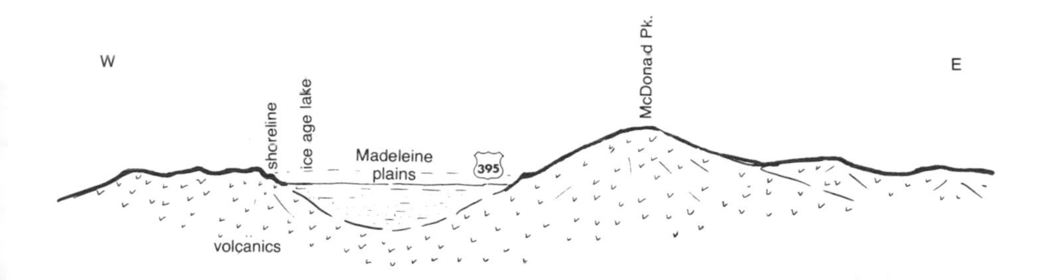

Section across U.S. 395 between Susanville and Alturas. The volcanic rocks must be resting on a basement of sierran granite.

ALTURAS — OREGON LINE
(40 miles)

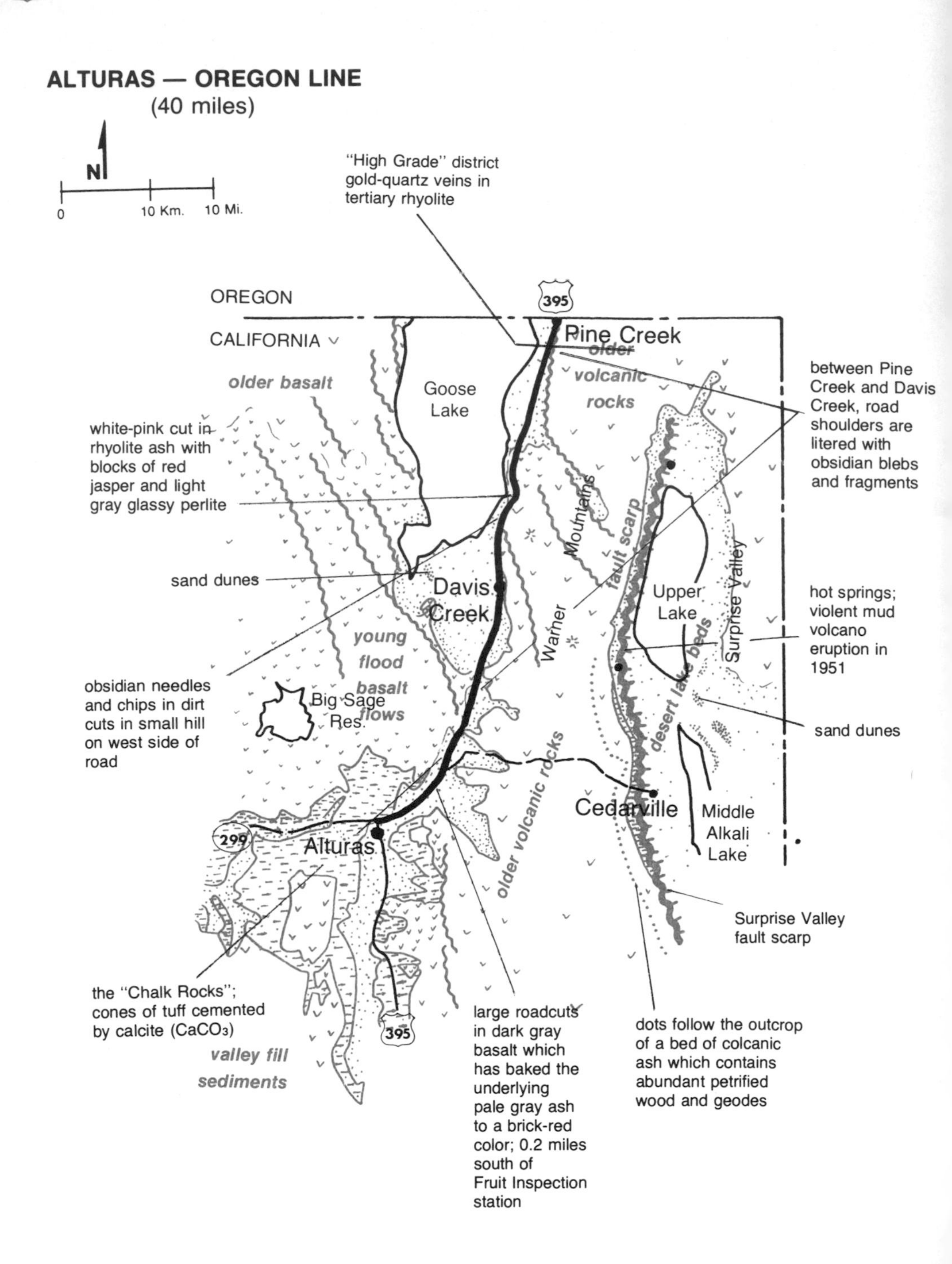

Basalt flow covering rhyolitic volcanic ash baked red where it was heated. Roadcut on U.S. 395 a few miles north of Alturas.

ALTURAS — OREGON LINE

All the rocks exposed along this road are volcanic, mostly either black basalt lava flows or very light-colored rhyolitic volcanic ash. Some were erupted during Miocene time, 15 to 30 million years ago while the lava plateau was forming and the others more recently as it broke up along faults to form the present landscape.

A raised block of the lava plateau, the Warner Mountains, forms the skyline east of the road. Lower hills west of the road are composed mostly of younger volcanic rocks erupted while the old lava plateau was breaking up along faults. The road follows a fault-block valley now partially filled with younger volcanic and sedimentary materials.

Section across U.S. 395 north of Alturas. Big fault blocks form the framework of the landscape.

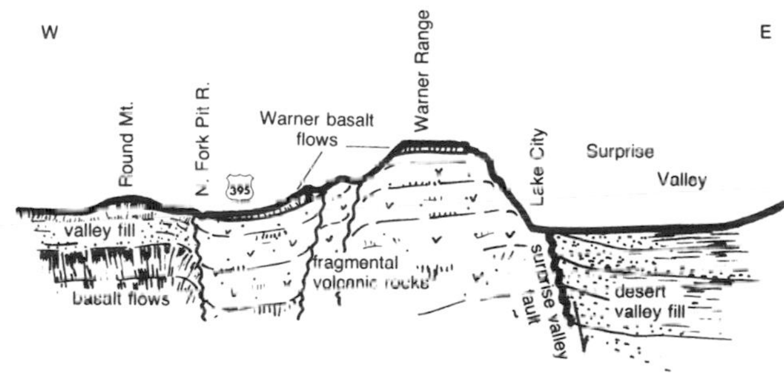

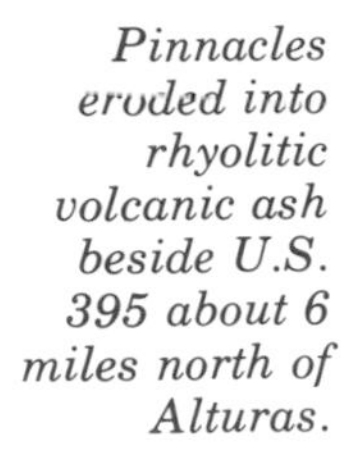

Pinnacles eroded into rhyolitic volcanic ash beside U.S. 395 about 6 miles north of Alturas.

For about 12 miles north of Alturas, U.S. 395 follows the North Fork of the Pit River, passing exposures of black basalt, white rhyolite volcanic ash and loose gravels, all part of the material filling the fault-block valley. Farther north the valley is more open and forms the basin for Goose Lake. The lake is much smaller now than it used to be because the climate is drier than it was during the last ice age and much of the remaining water supply is diverted to irrigation. Goose Lake has already shrunk enough that it no longer has an outlet and is becoming alkaline. Fish still survive and grow quite large but they have a terrible flavor.

Bubbly basalt lava flow. White crust on the rock is caliche, a deposit that forms in desert soils.

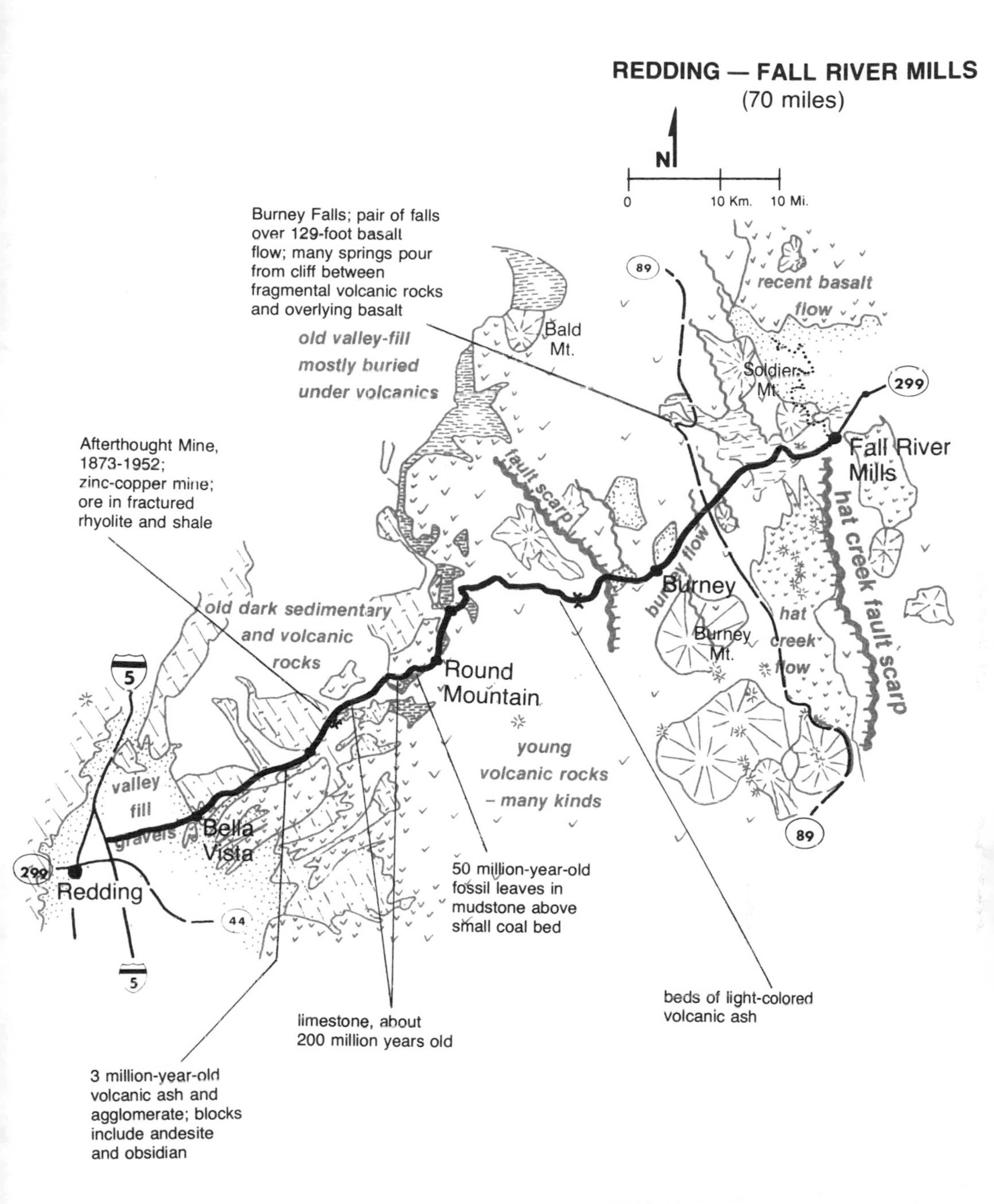
299
REDDING — FALL RIVER MILLS
(70 miles)
N
0
10 Km.
10 Mi.
Burney Falls; pair of falls over 129-foot basalt flow; many springs pour from cliff between fragmental volcanic rocks and overlying basalt
old valley-fill mostly buried under volcanics
Bald Mt.
89
recent basalt flow
Soldier Mt.
299
Fall River Mills
Afterthought Mine, 1873-1952; zinc-copper mine; ore in fractured rhyolite and shale
fault scarp
burney flow
Burney
hat creek fault scarp
old dark sedimentary and volcanic rocks
Burney Mt.
hat creek flow
5
Round Mountain
young volcanic rocks – many kinds
valley fill gravels
Bella Vista
89
299
Redding
44
5
50 million-year-old fossil leaves in mudstone above small coal bed
beds of light-colored volcanic ash
limestone, about 200 million years old
3 million-year-old volcanic ash and agglomerate; blocks include andesite and obsidian

california 299

REDDING — FALL RIVER MILLS

Redding is at the northernmost end of the Sacramento Valley where the thin edge of valley-fill sediments laps onto the torn southern margin of the Klamaths — the line of severance between them and the Sierra Nevada. California 299 crosses rolling, forested hills eroded into Sacramento Valley sediments from its junction with Interstate 5, north of Redding, to Bella Vista, about 7 miles east.

From Bella Vista east to Round Mountain, a distance of nearly 20 miles, the road follows the southern margin of the Klamaths, approximately tracing the boundary between them and the Cascades. Hills north of the road are underlain by older sedimentary rocks of the Klamaths whereas those to the south are composed mostly of Cascade volcanic rocks. Heavy vegetation and deep soils cover most of the hillslopes in this area making the bedrock difficult to see. But there are occasional roadcuts in dark rocks, platy looking because of their slaty fracture, that betray the underlying presence of the older metamorphic rocks of the Klamaths. Cascade volcanic rocks visible from the road are mostly brownish volcanic ash.

Blocks of weathered lava litter the surface of an old flow beside California 299 about 15 miles east of Redding.

Abandoned workings of the old Afterthought Mine,nowinactive, are conspicuous along the south bank of Cow Creek east of Bella Vista. This mine produced impressive amounts of copper and zinc during a period of many decades starting about the time of the Civil War and lasting until after World War II. The ore deposits occurred in volcanic rocks, light brownish-yellow ash deposits that are well-exposed in large roadcuts even though the mines across the creek are difficult to reach.

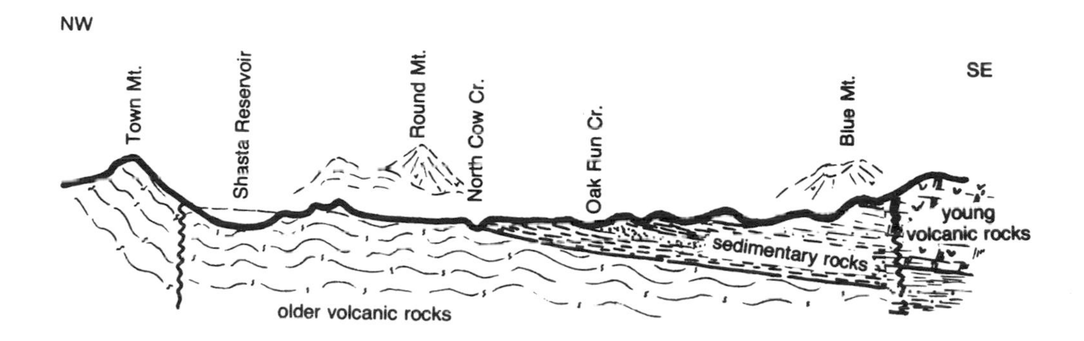

Section across California 299 between Bella Vista and Round Mountain. Younger sedimentary rocks of the Sacramento Valley lap onto folded volcanic rocks of the Klamaths – the kind exposed near the Afterthought Mine.

From Round Mountain to Fall River Mills, California 299 winds through a Cascades landscape composed of volcanic rocks erupted so recently that they have hardly been touched by erosion. The original volcanic landforms are preserved almost intact, quite a contrast to most other landscapes which are carved into the rocks by various processes of erosion. Streams are scarce in the Cascades because the fresh volcanic rocks are very porous and absorbent, leaving little surface runoff.

Burney Falls, a few miles north of the highway on California 89, is an interesting place. Some distance upstream from the falls, Burney Creek disappears underground into the pore spaces of a lava flow that fills its bed. The water continues to flow through the lava flow, hidden from view, until it begins to emerge again just upstream from the falls. Much of the water in the falls is actually pouring from the cut edge of the lava flow exposed in the face.

FALL RIVER MILLS — ALTURAS
(74 miles)

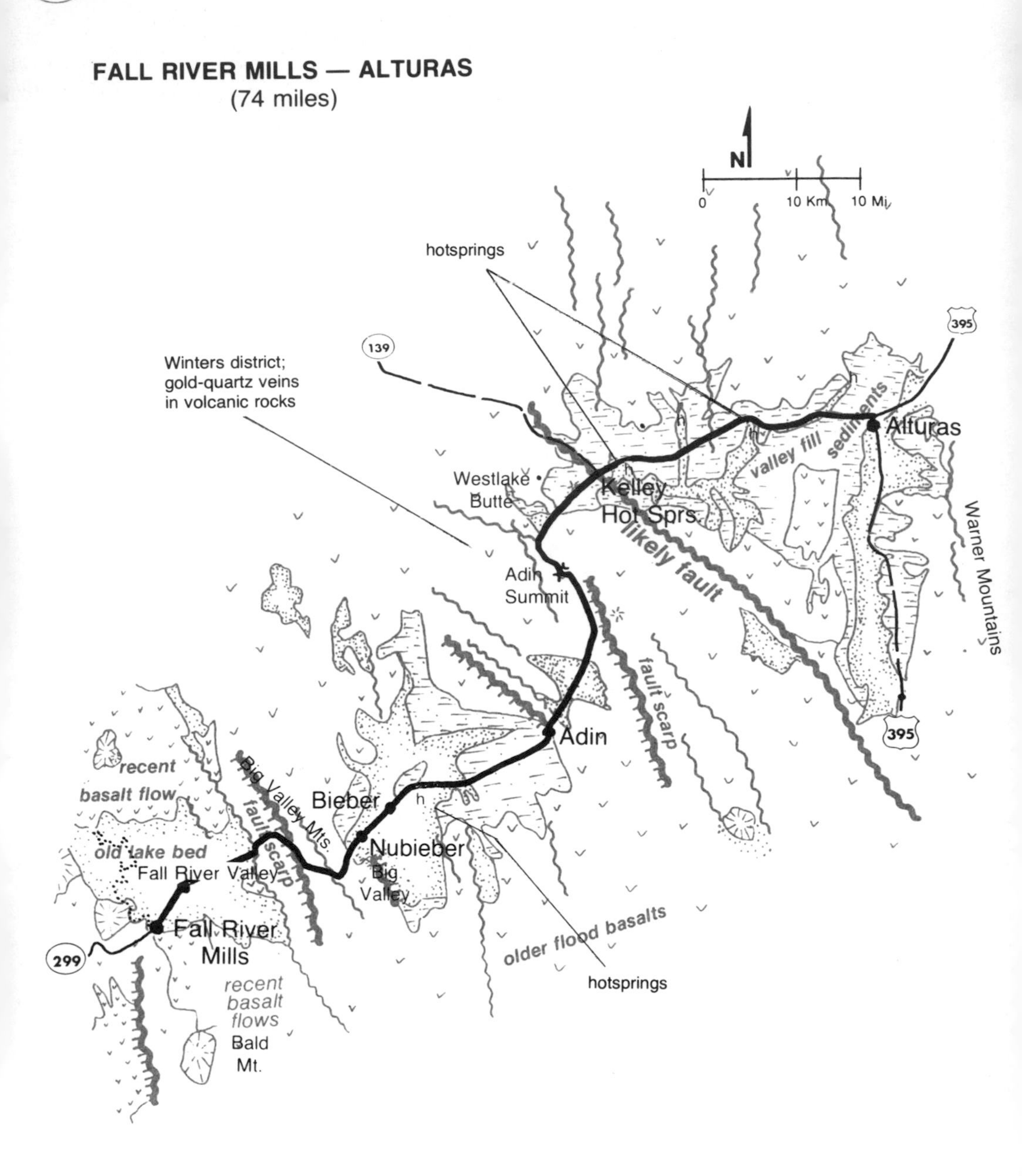

Volcanoes on the skyline, west across the Fall River Valley.

FALL RIVER MILLS — ALTURAS

The boundary between the Cascades and the Modoc Plateau lies just west of Fall River Mills. Between there and Alturas, California 299 crosses three large fault-block basins separated by two fault-block ranges — all part of the Modoc Plateau. Rocks exposed in the ranges are entirely volcanic — black basalt lava flows and occasionally volcanic ash beds mostly erupted before formation of the basins and ranges. Rocks exposed in the valley floors are partly sediments washed into them from the surrounding mountains and partly volcanic rocks erupted after formation of the basins and ranges.

Fall River Mills is at the west side of Fall River Valley and Adin is on the eastern edge of Big Valley. Both valleys are fault-block basins separated by the Big Valley Mountains, a fault-block range.

An enormous lava flow of black basalt, shown on most road maps, fills the whole northern end of the Fall River Valley. It was erupted sometime within the last few thousand years from one of the volcanoes in the Medicine Lake Highlands, about 40 miles north of Fall River Mills. Lava Beds National Monument includes the northeastern part of this area of recently intense eruptive activity where the volcanic rocks are still so fresh that they look almost as though they might have formed yesterday.

Fall River Springs, about 7 miles north of Fall River Mills, is one of the largest springs in the country. The water appears to be coming from Tule Lake and Clear Lake Reservoir, about 50 miles north. Large springs are fairly common in areas of volcanic rocks because many lava flows are very open and porous, providing natural avenues for flow of underground water.

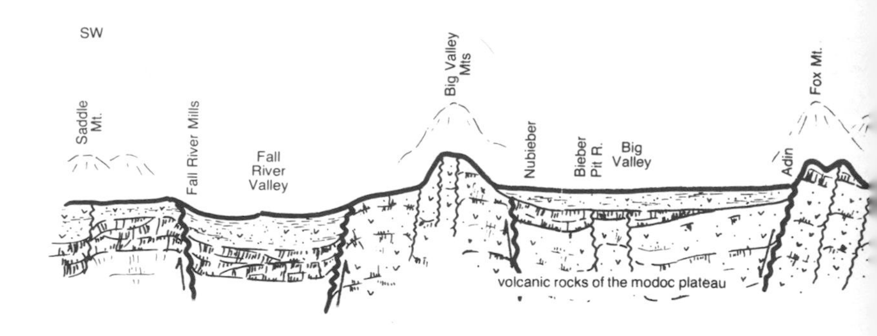

Sizeable lakes, perhaps impounded by lava flows, formerly existed in both the Fall River Valley and Big Valley. Extensive flat surfaces, north of the road in Fall River Valley and crossed by the road between Bieber and Nubieber in Big Valley, are deeply underlain by fine-grained lake sediments that blow in the wind during dry seasons and become very soggy when the weather is wet. More hilly areas of the valley floors are underlain by deposits of mud, sand, and gravel swept in from the surrounding mountains and now being eroded into hills by the Pit River and its tributaries. Large expanses of both valley floors are covered by lava flows but most of these are not visible from the road.

Tilted layers of white volcanic ash record a series of explosive eruptions on the Modoc Plateau.

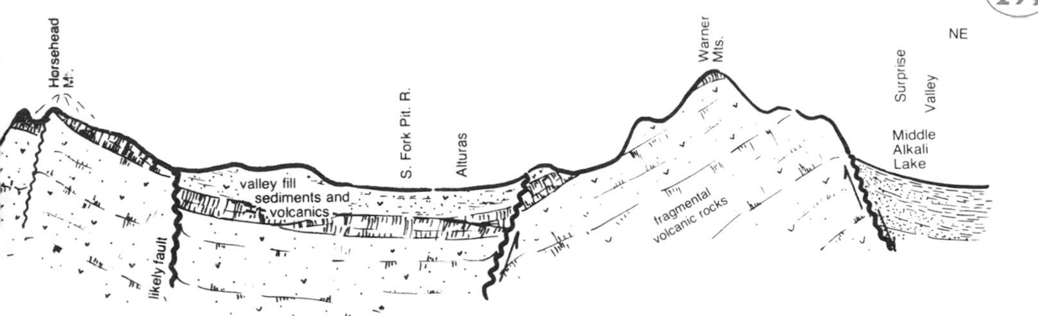

Section across the Modoc Plateau along the line of California 299. A thick sequence of volcanic rocks broken by faults into mountains and valleys forms the basis of the landscape.

Between Adin and Canby the road winds across another of the high mountain blocks uplifted by movement along faults in the Modoc Plateau during the last 15 million years. All the rocks exposed along the way are volcanic — lava flows of dense black basalt and thick beds of gray volcanic ash. These were erupted before the Modoc Plateau was broken by faults; most of them are probably between 20 and 30 million years old.

There are a few exposures of volcanic rock about midway between Canby and Alturas but most of the route is across deposits of loose sediment — mud, sand and gravel. Canby and Alturas are at opposite sides of one of the large basins let down along faults in the Modoc Plateau several million years ago. After the basin had formed, it was partially filled by loose sediments washed in from the surrounding mountains and by younger lava flows and volcanic ash falls. Then the Pit River eroded its valley into the basin-floor deposits, creating the roadside landscape we see today.

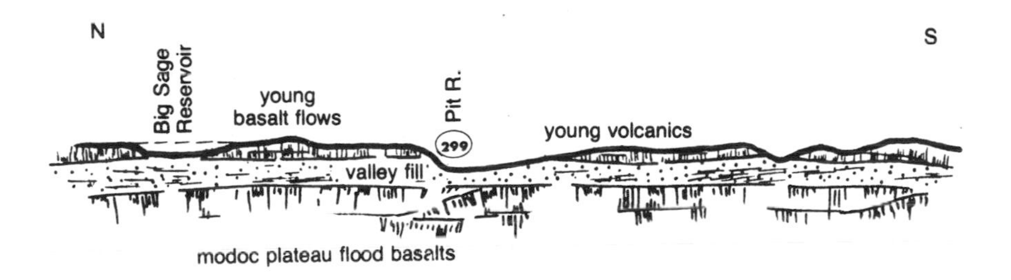

Section showing modern landscape eroded into valley-fill sediments and young volcanic rocks.

JUNCTION CALIF. 70 — LASSEN NATIONAL PARK
(100 miles)

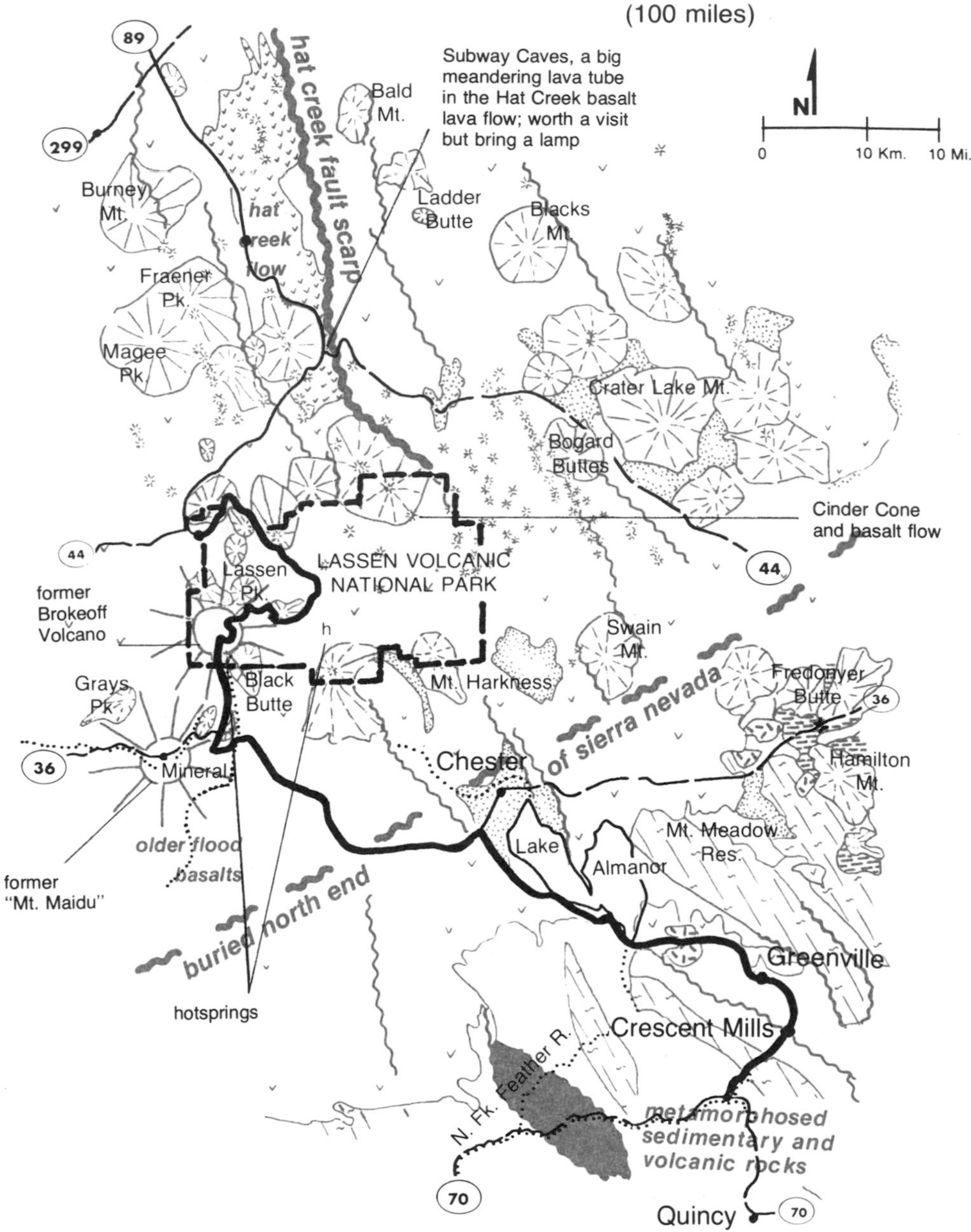

california 89

JUNCTION CALIFORNIA 70 — LASSEN PARK

Between its junction with Highway 70 and Lake Almanor, Highway 89 crosses the northern end of the Sierra Nevada. Rocks exposed along the way are old seafloor sediments now converted into slates by the heating and squeezing they received as they were jammed into the marginal trench.

Just north of Lake Almanor, which is a reservoir and not a natural lake, the road crosses the northern edge of the Sierra Nevada buried beneath thick deposits of young volcanic rocks. There is no hint of the boundary in the landscape, only a sudden change from deformed old sedimentary rocks to young volcanic rocks, and the knowledge that rocks similar to those south of Lake Almanor next appear about 60 miles to the west in the peaks of the Klamath Mountains that overlook the northern end of the Sacramento Valley.

Deposits of travertine built by a hot spring beside Highway 89 about a mile north of its junction with California 70.

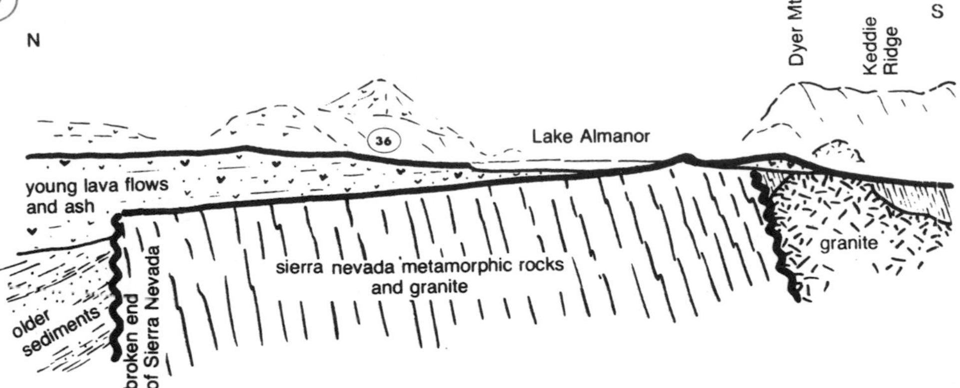

Section showing the broken end of the Sierra Nevada buried beneath young volcanic rocks just north of Lake Almanor. Sediments beneath the volcanics were deposited when the area was flooded by sea water after separation of the Klamaths and Sierra Nevada.

All the rocks exposed along California 89 between Lake Almanor and Lassen Park are volcanics erupted by Cascade volcanoes within the last few million years. Mt. Maidu, just west of California 89 near Morgan Summit is the wreck of a very large volcano. It collapsed about a million years ago to form a large subsidence crater after erupting several enormous lava flows.

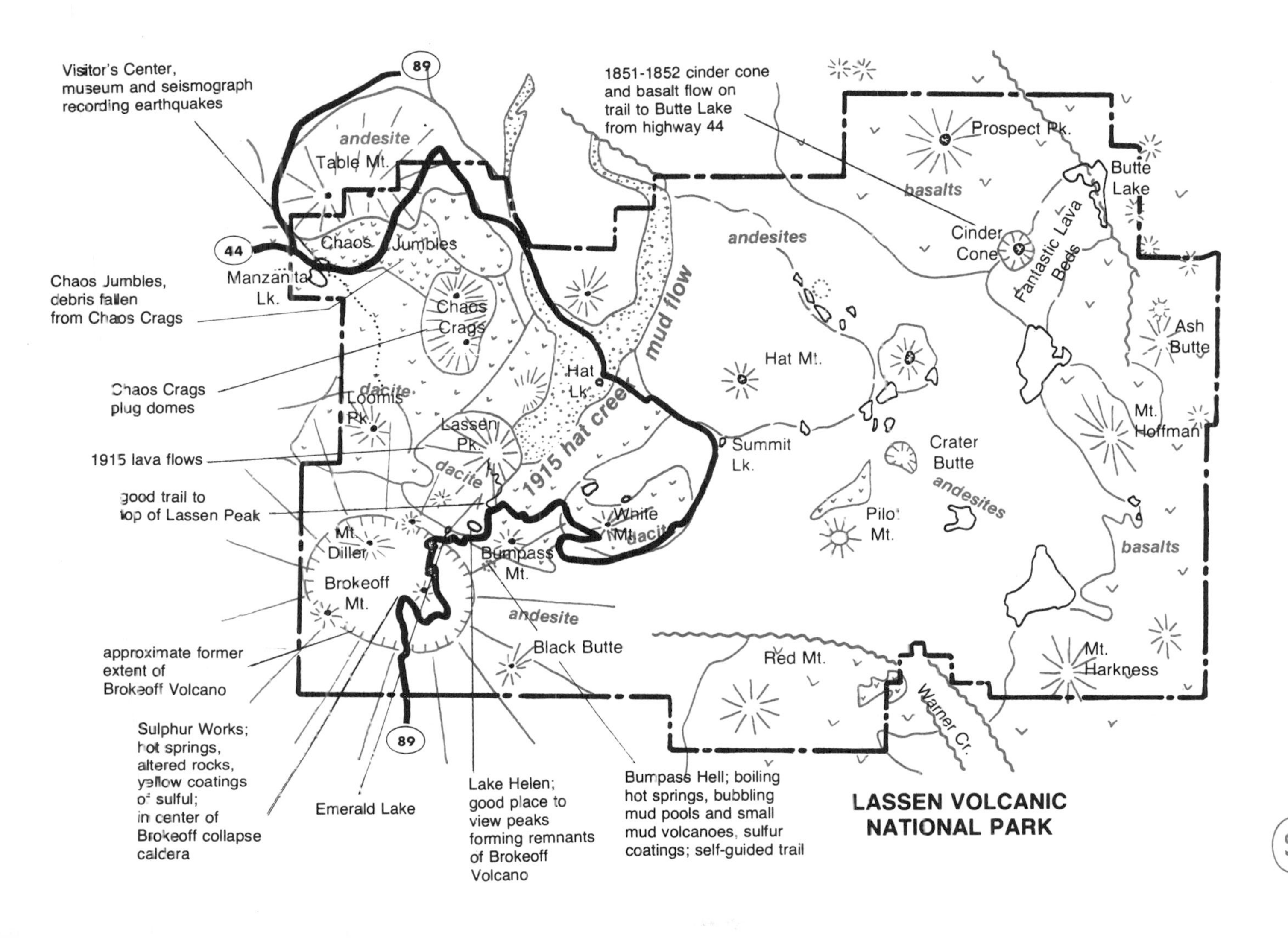
LASSEN VOLCANIC
NATIONAL PARK
89
44
Visitor's Center, museum and seismograph recording earthquakes
Chaos Jumbles, debris fallen from Chaos Crags
Chaos Crags plug domes
1915 lava flows
good trail to top of Lassen Peak
approximate former extent of Brokeoff Volcano
Sulphur Works; hot springs, altered rocks, yellow coatings of sulful; in center of Brokeoff collapse caldera
Emerald Lake
Lake Helen; good place to view peaks forming remnants of Brokeoff Volcano
Bumpass Hell; boiling hot springs, bubbling mud pools and small mud volcanoes, sulfur coatings; self-guided trail
1851-1852 cinder cone and basalt flow on trail to Butte Lake from highway 44
andesite
Table Mt.
Chaos
Jumbles
Manzanita
Lk.
Chaos
Crags
dacite
Loomis
Pk.
Lassen
Pk.
dacite
Mt.
Diller
Brokeoff
Mt.
Bumpass
Mt.
andesite
Black Butte
Hat
Lk.
1915 hat creek
mud flow
White
Mt.
dacite
andesites
Hat Mt.
Summit
Lk.
Crater
Butte
andesites
Pilot
Mt.
Red Mt.
Warner Cr.
Prospect Pk.
basalts
Cinder
Cone
Fantastic Lava
Beds
Butte
Lake
Ash
Butte
Mt.
Hoffman
basalts
Mt.
Harkness

lassen volcanic national park

Mt. Lassen, the southernmost of the big Cascade volcanoes, distinguished itself by going into violent action on the afternoon of May 30, 1914 to start a series of eruptions that culminated a year later. Most of the activity ended in 1917 but occasional small eruptions continued through 1921. Lassen has been completely quiet since then and shows no sign of activity today. Considering the type of volcano that it is, Lassen Peak itself seems most unlikely ever to erupt again.

Mt. Lassen rises above a high volcanic plateau erected during a complicated history of eruptive activity extending through the past several million years. Most of the evidence of early events is now buried beneath the debris of later eruptions which have built several large volcanoes within the park.

Brokeoff volcano, often called Mt. Tehama, was once the geologic centerpiece of the area. It was a giant volcano that dominated the southern part of the park for hundreds of millenia and then destroyed itself, not in violence but by collapsing softly into the earth as magma was withdrawn from beneath to feed the eruptions. All that remains of Brokeoff volcano today is a large subsidence crater where the peak once stood and a few remnants of its outer flanks; Brokeoff Mountain and Mount Diller are the largest of these.

Mt. Lassen rose after Brokeoff volcano sank. Most large volcanoes are built by a series of eruptions from a central vent but Mt. Lassen quite literally rose full grown from the earth. It is a plug dome, a bulging protuberance of extremely viscous magma that pushed up through the surface as though it were an enormous puffball mushroom growing after a spring rain. Such volcanoes rise, steaming and popping blocks of rock off their outer shells, within a few years time and then solidify where they stand to become ragged lumps of solid lava clothed in sliding screes of broken rock. Lassen is one of the largest of such volcanoes known and one of the few to erupt again after it had stopped growing.

Lassen's recent activity seems to have consisted mostly of steam explosions probably caused by surface water seeping downward through cracks to the vicinity of a large mass of hot rock hidden deep within the mountain. Ash blown from the volcano during those eruptions consists mostly of broken fragments of rock ripped from the solid interior of the mountain and blasted out with the escaping steam. The craters in the top of Mt. Lassen were blasted out of the solid rock and are not the vents from which material was erupted to build the mountain, most plug dome volcanoes have no craters in their summits.

During mid-May of 1915, a red glow appeared in the sky above Mt. Lassen and globs of glowing red lava were blown out of the crater to roll down the slopes below. Lava overflowed the rim of the crater for a few hours sending one small flow pouring slowly down the western slope of the volcano. Visitors see it today as a black tongue extending a few hundred feet downhill from the crater.

About the time the lava flow was slowly poking its tentative tongue down the western slope of the peak there was violent activity on the northeastern flank where a devastating mudflow overwhelmed long stretches of the Lost Creek and Hat Creek valleys. It started on a cloudy night so no one actually saw what happened but some sort of blast of steam and water levelled a large tract of forest and created an enormous quantity of mud. Because it is much heavier than water, mud exerts a much greater buoyant effect and large mudflows move huge boulders as though they were marshmallows. Several ranches were completely buried under deposits of mud, and littered with boulders weighing as much as 20 tons, all within a few minutes. The fact that some of the boulders were red hot lava added considerable interest to the occasion. Mudflows often accompany volcanic eruptions and usually, as at Lassen, they are more dangerous and destructive than the lava flows.

On May 22, 1915, the day after the lava and mud flows, Mt. Lassen blew its top in an astounding explosion that blasted a black mushroom cloud of steam and rock fragments at least 25,000 feet into the sky, casting a pall over the surrounding countryside. The cloud was seen from the cupola of the state capital in Sacramento and there was panic in the streets of Redding. After this superb effort, the volcano subsided and subsequent activity consisted only of small steam explosions. The eruptions left the mountain a battered wreck, its sides streaked with black stripes left by the big mud flows. These are now so well covered by new vegetation that almost no visible trace remains of the big eruption.

Chaos Jumbles, foreground, and Chaos Crugs, background.

Chaos Crags, a group of ragged hills near the north side of the park, are another plug dome much smaller and younger than Lassen but basically similar. Reliable people reported that they were still steaming in 1850 and geologists figure that they rose sometime around 1700. Future eruptions in Lassen Park will more likely involve the protrusion of more plug domes similar to Chaos Crags than renewed activity of Lassen itself.

About the time the mass of molten magma that formed Chaos Crags had protruded to its present height, blasts of steam blew a crater into the base of the mountain, undermining its northwestern side. The undercut slope collapsed, spilling a mass of broken rock over nearly a square mile of land to create Chaos Jumbles, a rolling sea of angular chunks of glossy brown rock dumped in a matter of seconds. Few lichens grow on the fresh surfaces of the blocks, nor does much shrubbery grow among them — evidence that the rockfall happened within the last few centuries. Geologists have puzzled for years over the question of why rockfalls such as Chaos Jumbles spread out over a large area instead of simply tumbling to the base of the mountain in a heap. One theory holds that they ride across the landscape on a trapped cushion of compressed air and another that the broken chunks keep themselves in suspension for a few seconds by banging into each other as the entire mass moves. Recent moon photographs taken by astronauts show features that look very much like rockfalls — support for the second theory because the moon is an airless planet.

Andesite agglomerate, a disorderly mixture of large and small angular chunks all dumped together. The big pieces here are about the size of grapefruit.

All the rocks in Lassen Park are volcanic and nearly all those likely to be noticed by visitors are varieties of andesite and dacite, the commonest kinds of rock in volcanic chains such as the Cascades. Andesites come in a variety of colors, usually some medium shade of gray, brown, or reddish brown, but are never either white or black. Dacites are generally similar but somewhat lighter in color. They both come in a variety of forms and disguises depending upon what the volcano happened to be doing when they were erupted. Agglomerates are very common, they are messy mixtures of large and small fragments coughed out of the vent and settled on the slope of the volcano, or carried from its flanks by mudflows. If all the fragments are very small, as is usually the case in deposits that settled more than a mile from the vent, the rock is called an ash bed. Solid lava is also exposed in the park, most conspicuously in the vicinity of Lake Helen and Emerald Lake. Andesite and dacite lavas generally contain small crystals peppered through a smooth matrix of very fine-grained rock.

Lava erupted from Lassen in 1915 is famous among geologists because it contains both light and dark-colored rock in streaks and blotches as though it were an attempt at some sort of volcanic marble cake. The explanation for this has become the subject of a considerable literature which has not yet produced any consensus of opinion. Some geologists believe that these streaky rocks formed as an originally homogeneous magma began to separate into two different kinds of rocks, and others that they fomed through incomplete mixing of originally different magmas. All participants to the dispute seem agreed that the question of how these rocks should be interpreted is very important even though they can't agree on the answer.

Close-view of an andesite. Crystals of white feldspar about the size of peanuts are scattered through a fine-grained matrix of smooth brown rock.

california 89 lassen park road

It is difficult to get a clear impression of the surrounding geology from the road through Lassen Park. The area does not contain a single volcano but an entire complex of them active in various ways for a long time. Such complexity compounded of unfocused late activity superimposed upon the wreckage of the volcanic past is typical of many large volcanoes and adds greatly to their charm once it is understood.

Between the southern entrance station and Emerald Lake, the road winds through the remains of Brokeoff volcano (Mt. Tehama) without ever coming to a good vantage point from which the whole story can be visualized. The basin created by subsidence of the old peak is west of the road and hard to see because it is very large and because later volcanic activity has largely blurred its outlines.

Section through Lassen Park showing the complex array of volcanic rocks developed through several million years of eruptions.

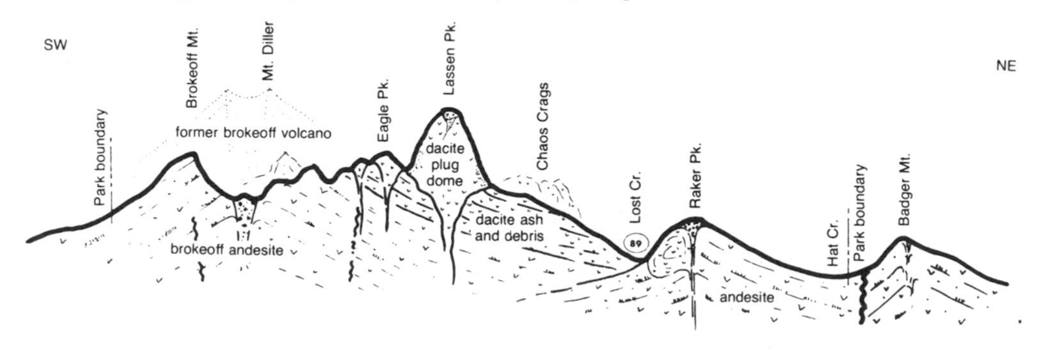

View north up Little Hot Springs Valley to Lassen peak.

Bumpass Hell and Supan Springs at the Sulphur Works are the two groups of volcanic hot springs and gas vents seen by most visitors to the park. They are associated with Brokeoff volcano, rather than with Mt. Lassen. Surface water seeps downward through cracks to the vicinity of large bodies of very hot rock that may be only a few hundred feet underground. There the water is heated, reacts with the volcanic rocks, and boils back to the surface as mineralized hot water or steam. Much of the dissolved mineral matter is deposited around the springs where the water begins to cool, creating fantastic structures of siliceous sinter and white travertine. Geologists are fascinated by such hot springs because it seems likely that circulating hot waters such as these are responsible for emplacement of many valuable ore deposits.

Emerald Lake, and the other lakes beside the road where it passes through the higher parts of the park, is the result of glacial erosion and not of volcanic activity. During the last ice age much of Lassen Park was covered by ice and snow and large glaciers poured down the peaks to scour the valleys below. Beautiful exposures of solid lava are beside the road where it winds through this highest part of the park.

From Hat Lake northward for several miles the road follows the valley of Lost Creek, one of those devastated by the mudflow of May, 1915. The sight no longer appalls park visitors as it did a few decades ago because a new growth of vegetation has almost covered the mudflow. Occasional specimens of the streaky mixed lava erupted in 1915 are strewn along this creek. They are easiest to find around the Devastated Area parking lot.

Chaos Jumbles and Chaos Crags are fascinating sights at the north end of Lassen Park; it would be hard to find better views of either a plug dome or a rockfall anywhere. Manzanita Lake at the north end of the park exists because the Chaos Jumbles rockfall dammed Manzanita Creek.

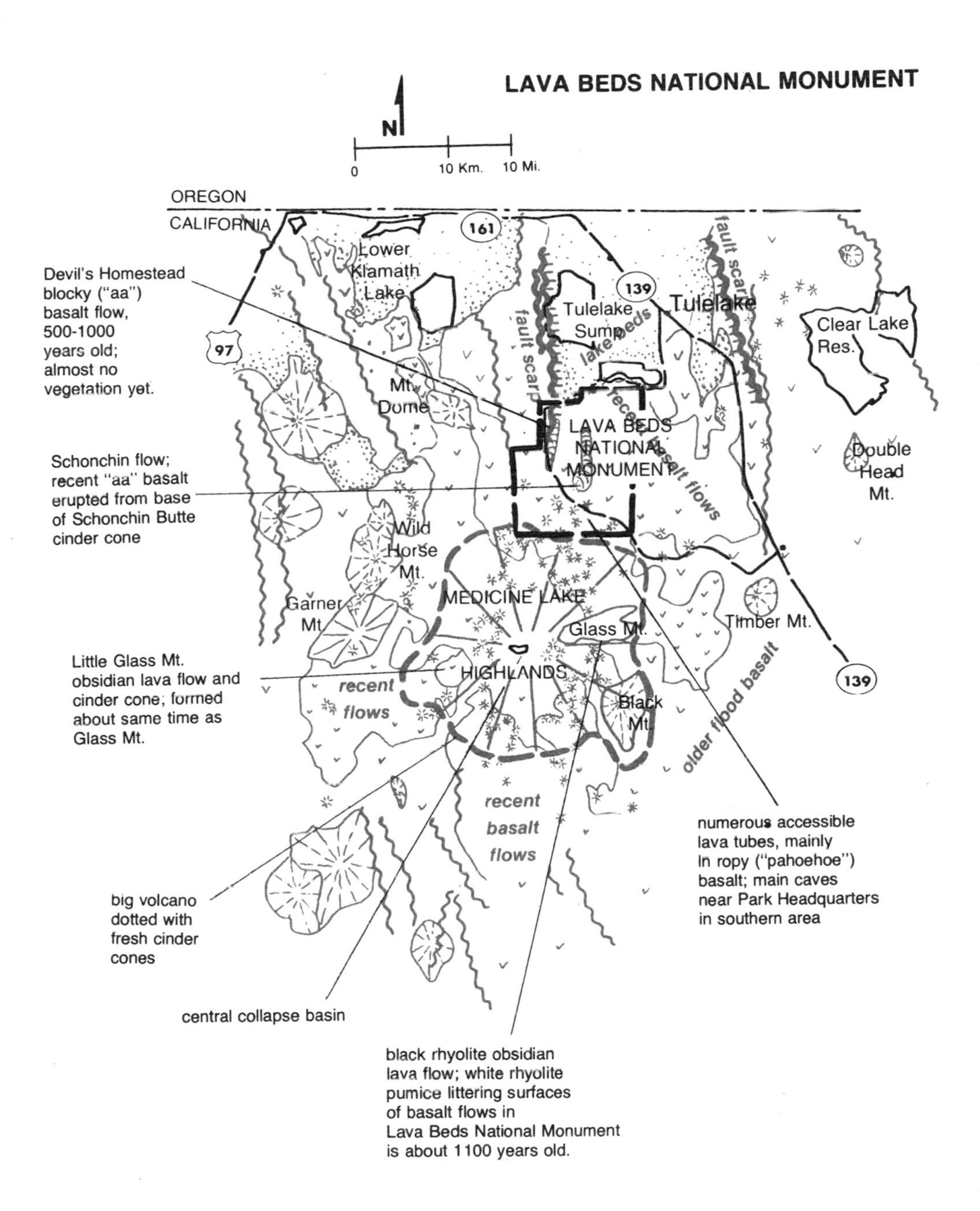
LAVA BEDS NATIONAL MONUMENT
N
0
10 Km.
10 Mi.
OREGON
CALIFORNIA
Devil's Homestead blocky ("aa") basalt flow, 500-1000 years old; almost no vegetation yet.
Lower Klamath Lake
161
139
97
Tulelake Sump
Tulelake
fault scarp
fault scarp
lake beds
Clear Lake Res.
Mt. Dome
LAVA BEDS NATIONAL MONUMENT
recent basalt flows
Double Head Mt.
Schonchin flow; recent "aa" basalt erupted from base of Schonchin Butte cinder cone
Wild Horse Mt.
Garner Mt
MEDICINE LAKE
Glass Mt.
Timber Mt.
HIGHLANDS
Black Mt.
139
Little Glass Mt. obsidian lava flow and cinder cone; formed about same time as Glass Mt.
recent flows
older flood basalt
recent basalt flows
numerous accessible lava tubes, mainly in ropy ("pahoehoe") basalt; main caves near Park Headquarters in southern area
big volcano dotted with fresh cinder cones
central collapse basin
black rhyolite obsidian lava flow; white rhyolite pumice littering surfaces of basalt flows in Lava Beds National Monument is about 1100 years old.

Red hot basalt dribbled like candle wax to make this fantastic dripstone surface in the mouth of a spatter cone. Area shown is about 2 feet across.

lava beds national monument

A black landscape of basalt lava flows so freshly erupted that they look like something that belongs more to the moon than to the earth. That is Lava Beds National Monument where the volcanic rocks are so fresh and unweathered that visitors half expect to see them steaming and hear the noises of eruption. It is surprising to find that most of these rocks are actually more than 1000 years old and a bit sobering when we see how little the slow processes of soil formation have accomplished in that time. Soil forms slowly and is not a renewable resource within spans of time that have any human meaning.

All the flows in Lava Beds National Monument are basalt, the commonest volcanic rock. Basalt lava flows form the bedrock floors of the oceans, 70 percent of the earth's surface, and are common almost everywhere volcanoes erupt. Most of the specimens brought from the moon by the astronauts turned out to be common black basalt hardly distinguishable from that on the Modoc Plateau.

Molten basalt comes from deep beneath the crust from the mantle where the rocks are hot enough to melt but ordinarily remain solid because their position within the earth places them under tremendous pressure. The continental crust beneath this region has been stretched during the last several million years, causing it to break into large blocks that rose and fell alternately to become mountain ranges and basins. Pressure on the mantle beneath was reduced enough to generate molten basalt magma which rose to the surface along the same fractures that define the mountain ranges and basins.

Ropy surface formed as molten lava flowed beneath a thin crust of solidifying rock.

Even though molten basalt is the liquid that forms when melting begins in the mantle, all our evidence indicates that the mantle is actually composed of rocks heavier than basalt and different in composition. Extraction of a liquid having one composition from a melting solid with another is a familiar process to anyone who has ever sucked the sweet juice out of a thawing popsicle leaving white and tasteless ice on the stick. Similarly, basalt magma is the "juice" that forms when the mantle begins to melt.

Molten basalt is usually very fluid so it runs freely over the surface making big lava flows that resemble black puddles when seen from the air. Steam bubbles escape from the fluid lava quite freely so most basalt eruptions are quiet affairs by volcanic standards — volcanic violence is almost always caused by trapped steam.

Many basalt lava flows have rough surfaces covered by several feet of broken clinkers of lava, all filled with bubbles formed by escaping steam. Surfaces of these flows are forbidding expanses of boot-shredding angular rubble almost impossible to walk across. Other basalt flows develop surfaces almost smooth enough to invite bicycling even though they generally have a somewhat ropy texture. Despite their great differences in appearance, both these kinds of basalt lava flows are composed of exactly the same kind of rock. We often find both types of surface on different parts of the same flow and the reasons for the difference are not well understood.

Most small basalt eruptions, like those at Lava Beds National Monument, begin as fragments of molten magma are coughed out of a new vent by escaping steam. These cool in the air and the smaller ones drift downwind as a cloud of dark ash while the larger ones settle around the vent to build a loose heap called a cinder cone. After this activity has continued for a while, anywhere from several days to several years a lava flow bursts through the base of the cinder cone and pours over an area that may be as small as a city block or as large as several square miles. Most cinder cones then become quiet and very few ever erupt again. They tend to be single-shot volcanoes.

Basalt lava flows do a lot of interesting things in their few days or weeks of mobility before they finally chill into rigid rock. At first the lava pours slowly over the countryside advancing at a rate usually measureable in terms of a few feet per hour, rarely as fast as a person can walk. Then things get more complicated as the surface of the flow cools enough to solidify while the interior remains molten and fluid. Continued flowage under the solid surface tears it open in places, exposing the molten lava beneath, and jams it tight in others making pressure ridges resembling those that form in moving pack ice. Occasionally molten basalt squeezes up through cracks in the surface making strange extrusions that look as though someone might have stepped on an enormous tube of black toothpaste.

Gaping holes appear when the thin roofs covering lava caves begin to collapse.

Roof collapses let the sun into a lava cave. A stronger portion of the roof still stands forming a natural bridge.

Large caverns form within the lava flow as molten basalt drains from beneath the solid surface crust. These are often lined with fantastic little dripstone constructions formed as molten basalt dribbled off the ceiling and down the walls as it cooled. Many basalt caves have horizontal ledges along their walls recording periods when the lava surface remained stationary long enough to cool and solidify a bit before draining away some more. Eventually the roofs of basalt caves collapse forming sinkholes and long, winding depressions on the surface of the flow are crossed by natural bridges formed by remnants of the original roof.

Some of the deeper lava caves have permanent ice in their lower parts. Dense cold air sinks into the bottom of the cave in winter and will remain there until it is displaced by even colder air sinking down from above, so the only new air entering a poorly ventilated cave is colder than that already there. Rocks are excellent insulators so a large mass of cold air and ice can easily survive a long summer provided there is no way for warm air to blow through. Ice in the bottoms of caves is not a relic of the glacial periods, as many people assume, but simply the remains of the cold days of winter.

A series of small cinder cones forms the skyline in this view of the Medicine Lake highland.

Medicine Lake, just south of Lava Beds National Monument, is in the middle of another fascinating volcanic center. A long series of eruptions built a large shield volcano which had very gently sloping sides. Its summit eventually collapsed as the reservoir of molten magma beneath emptied, creating an oval caldera about 6 miles long, 4 miles wide, and 500 feet deep. Later eruptions built a ring of eight smaller volcanoes around the rim of the basin nearly obscuring it. Medicine Lake occupies all that is left.

Some of the later eruptions in the Medicine Lake area produced rhyolite, a light-colored volcanic rock that behaves very differently from basalt. Molten rhyolite magmas have a stiff and pasty consistency closely resembling that of modelling clay. When such a magma contains a lot of steam, as some of them do, it may blow off in violent explosions that cover the countryside for miles around with thick deposits of light colored ash and chunks of frothy rhyolite glass called pumice — so full of gas bubbles that it floats on water. Some of these deposits in the Medicine Lake area are mined

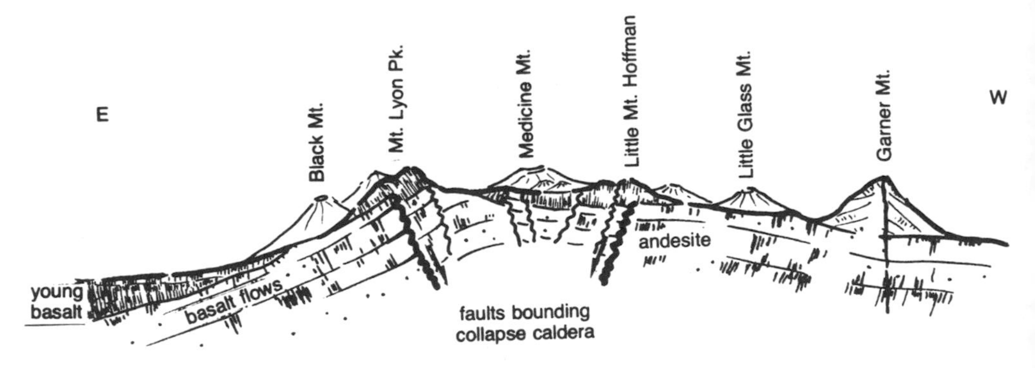

Section across the Medicine Lake highland showing young volcanoes built on the collapsed remains of an older one that foundered as lava was withdrawn from beneath.

Angular chunks of feather-weight white pumice littering the landscape of Lava Beds Monument were produced during the latest eruptions in the Medicine Lake highland. The larger of these are about the size of acorns.

for light-weight aggregate and portable garden rocks. Rhyolite magmas that happen not to contain much steam may extrude quietly through the surface where they make bulging lava domes and swollen lava flows composed mostly of obsidian, a black volcanic glass. Unlike basalt, which is also black, obsidian is very shiny and breaks into slick fragments with razor sharp edges. There are a number of obsidian lava flows in the Medicine Lake area, especially around Mount Hoffman about 2 miles east of the lake.

CROSS-REFERENCES

More about the nature and origin of andesite and related volcanic materials in California 89: Junction 70 — Lassen Park.

More about Mount Shasta in Chapter V, Interstate 5: Dunsmuir — Oregon line.

More about basin and range faulting in the introduction to Chapter V, The Cascades and Modoc Plateau.

glossary

We have tried to avoid using technical terms but a few slipped in anyway. The glossary also contains a few words that we didn't use ourselves but often found in books and articles intended for non-geologists.

Andesite. A volcanic rock intermediate in composition between rhyolite and basalt. Andesites have about the same composition as common granite. Most of them are brown, red or gray although some are green. Andesites are common in the Cascades.

Ash. Small fragments of rock or globs of magma blown from a volcanic vent by escaping steam.

Auriferous. "Gold-bearing," the word has an antique sound but is still used occasionally in referring to the dry gravels of the Sierra Nevada.

Basalt. An absolutely black volcanic rock most often seen as lava flows but may also form beds of black volcanic cinders. Basalts often become barn red when exposed to the weather. By far the commonest volcanic rock, basalts are abundant wherever volcanoes have been active.

Batholith. A very large mass of granite, one having an outcrop area greater than about 40 square miles. Smaller masses of granite are called "stocks."

Blueschist. An unusual kind of metamorphic rock that may contain any of several blue minerals distributed through a matrix of black minerals. Blueschist chunks are distinctly heavier than most other rocks and hard to break with a hammer.

Biotite. Black mica, one of the common minerals in igneous and metamorphic rocks. Individual crystal grains are shaped like flat flakes.

Calcite. The mineral that forms limestones and also occurs in many other kinds of rocks. It is calcium carbonate, a substance familiar in seashells and boiler scale.

Caldera. A large crater formed by collapse of a volcano.

Cinder Cone. A type of small basaltic volcano that consists of a pile of cinders several tons to hundreds of feet high and a few lava flows. Most cinder cones are one-shot volcanoes that erupt for a few weeks and then become permanently extinct.

Cinnabar. Mercury sulfide, the common ore of mercury. It is bright red.

Composite Volcano. A volcano composed of both volcanic ash and lava flows. Most composite volcanoes are large, symmetrical cones. Shasta is a typical example.

Dacite. A light-colored volcanic rock about midway between andesite and rhyolite in composition and appearance.

Eclogite. An unusual kind of rock found in many serpentinites along with chunks of blueschist and jade. Eclogites are heavy, dark-green, jadeite-bearing rocks flecked with red crystals of garnet.

Fault. A fracture in the earth's crust separating two blocks that shift past each other. Rocks on opposite sides of the fault do not match.

Feldspar. The most abundant mineral in many igneous and metamorphic rocks. There are many varieties of feldspar. The commonest are salmon pink, white, or greenish white. All are translucent or chalky and break along flat surfaces.

Flood Basalt. An enormous basalt lava flow; most flood basalts cover hundreds of square miles and are more than 50 feet thick. A series of such lava flows forms a high plateau instead of a volcanic mountain.

Franciscan Rocks. The name applied collectively to the crumpled sea floor sediments that form the bulk of the Coast Range. The name comes from the fact that these rocks were first closely studied around San Francisco. Often called Franciscan schists.

Fumarole. A volcanic gas vent.

Granite. An igneous rock composed mostly of clear quartz and white or pink feldspar. Most granites also contain a black mineral, either mica or hornblende. Crystal grains are usually rice-size or larger. In outcrops, granite is a gray or pink rock liberally peppered with black grains.

Great Valley Sequence. In this book the term is used to refer to the thick section of sea floor sediments that dip from the east flank of the Coast Range under the Great Valley. The same term has been more widely applied elsewhere — its meaning is somewhat confused but it is too deeply embedded in the geological vernacular of California to be abandoned.

Greenstone. A dark-colored volcanic rock that has been metamorphosed by prolonged heating. Greenstones are usually rather messy and nondescript in all respects except for their color which really is green.

Hornblende. A shiny black mineral common in many igneous and metamorphic rocks. Well-formed crystals are shaped about like stubby pencils.

Ice Age. A climatic episode in which very large glaciers develop, probably more a matter of heavy snowfall than of coldness. There have been at least four major ice ages during the past 3 million years.

Inclusions. Odd fragments of one kind of rock suspended in another kind.

Knocker. A term often used in California to refer to chunks of blueschist, eclogite or jade weathering as knobs in their enclosing serpentinite.

Laterite. Red and yellow soils that form on deeply weathered rock in regions having very wet climates.

Lava. Molten rock, magma, that has been erupted from a volcano.

Limestone. A sedimentary rock composed mostly of calcite, the mineral that makes sea shells.

Lode. A deposit of valuable minerals still enclosed iin the original bedrock.

Magma. Molten rock.

Mantle. The largest part of the earth, a shell of black rock thousands of miles thick that underlies the continents and ocean floors and continues downward to the core.

Metamorphic Rocks. Rocks so changed by prolonged heating and deformation that they no longer resemble the igneous or sedimentary rocks they once were.

Opal. A non-crystalline variety of quartz.

Peridotite. A heavy, black igneous rock, one of the kinds that form the earth's mantle below the sea floor. Many of the rocks we have called "old sea floor and serpentinite" in this book are more technically known as peridotite.

Placer. A deposit of heavy minerals concentrated in stream gravels. Because they are heavy, gold and platinum commonly occur in placer deposits.

Plate. A portion of the earth's crust that moves as a unit as though it were rigid. Most of the rocks in northern California were created by collision of the North American and Farallon plates.

Plug Dome. A kind of volcano formed by protrusion of a mass of extremely stiff and pasty magma through the surface. Lassen is an unusually large plug dome.

Quartz. A very common mineral, it occurs in most kinds of rock in many different forms. Commonest as small, glassy-looking grains but agate, chert, jasper and rock crystal are all varieties of quartz. Composed of silica (silicon dioxide).

Rhyolite. A very light-colored volcanic rock most often seen as beds of volcanic ash. Comes in various pale shades of gray, yellow, pink and tan. Rhyolites correspond in composition to some of the lighter varieties of granite.

Richter Magnitude. A scale for measuring the sizes of earthquakes in terms of the amount of energy they release. Each unit jump on the Richter scale represents a tenfold difference in energy release.

Salinian Block. The part of the earth's crust that is moving northward along the west side of the San Andreas fault.

Seismograph. An instrument that detects and records vibrations of the earth.

Silica. Another name for quartz, silicon dioxide.

Silica-carbonate Rock. A rock composed of quartz and various carbonate minerals that is formed by alteration of serpentinite by circulating steam. It is common around the mercury mines in the Coast Range where it forms prominent, light-colored outcrops.

Silicious Sinter. A deposit of porous quartz laid down around the mouth of a hot spring or volcanic gas vent.

Slate. Mudstone that has become hard and developed a set of closely-spaced parallel fractures as a result of having been cooked and squeezed.

Subduction. A term often used to refer to the descent of old sea floor into the earth's interior.

Tailings. Waste rock dumped from a mill or placer mining operation.

Terrace. A flat or gently sloping bench, usually a remnant of an old river floodplain or of an old coast line. River terraces are conspicuous along most California streams and marine terraces along most of the coast.

Till. Sediment deposited by glacial ice, generally a disorderly mixture of unsorted mud, sand, gravel and boulders.

Travertine. A deposit of calcite laid down in a cave or around the mouth of a hot spring.

Unconsolidated. Loose sedimentary material not welded together into a solid rock. Sand is an unconsolidated sediment and sandstone is the corresponding rock.

Vein. Any mineral deposit that fills a fracture. Veins occasionally contain valuable minerals but most do not.

Volcanic Ash. Small fragments of solid rock or molten magma blown from a volcano. Ash settles downwind to make layered deposits that resemble sedimentary rocks.

Weathering. The various surface processes that decompose solid rock, turning it into soil.

selected references

These are only a minute fraction of the numerous articles and books that contain good information about the geology of northern California. Each of them contains references to other good literature.

Atwater, T., 1970, Implications of plate tectonics for the Cenzoic tectonic evolution of western North America, Geological Society of America Bulletin, vol. 81, p. 3513-3536.

Bailey, E.H., M.C. Blake, Jr., and D.L. Jones, 1970, On-land Mesozoic oceanic crust in California coast ranges, U.S. Geological Survey Professional Paper 700-C, p. 70-81.

Bailey, E.H., W.P. Irwin, and D.L. Jones, 1964, Franciscan and related rocks, Calif. Division of Mines and Geology, Bulletin 183, 177 p.

Bailey, E.H., editor, 1966, Geology of northern California, Calif. Division of Mines and Geology, Bulletin 190, 507 p.

Barbat, W.F., 1971, Megatectonics of the Coast Ranges, California, Geological Society of America, Bulletin, vol. 82, p. 1541-1562.

Bowen, O.E., Jr., editor, 1962, Geologic guide to the gas and oil fields of northern California, Calif. Division of Mines and Geology, Bulletin 181, 412 p.

Dickinson, W.R., and A. Grantz, editors, 1968, Proceedings of the conference on geologic problems of San Andreas fault system, Stanford University Publications in Geological Sci., vol. 11, 260 p.

Geological Map of California (Olaf P. Jenkins edition, 1958-1969), Calif. Division of Mines and Geology, Scale 1:25,000.

Hamilton, W.B., 1969, California and the underflow of Pacific mantle, Geological Society of America Bulletin, vol. 80, p. 2409-2430.

Irwin, W.P., 1960, Geologic reconnaissance of the northern Coast Ranges and Klamath Mountains, California, Calif. Division of Mines Bulletin 179, 80 p.

Maxwell, J.C., 1974, Anatomy of an orogen, Geological Society of America Bulletin, vol. 85, p. 1195-1204.

Wahrhaftig, C., 1960, A walker's guide to the geology of San Francisco, Mineral Information Service Supplement, Nov. 1966, 32 p.

Williams, H., 1929, Geology of the Marysville Buttes, Calif. Univ. Dept. of Geological Sciences Bulletin, vol. 18, p. 103-220.

Williams, H., 1932, Geology of the Lassen Volcanic National Park, California, Calif. Univ. Dept. of Geological Sciences Bulletin, vol. 21, p. 195-385.

Index